Stem Cell Therapy

Stem Cell Therapy

Practical Considerations

HALA M. GABR
Department of Clinical Pathology, Cairo University, Cairo, Egypt

WAEL ABO EL-KHEIR
Egyptian Society for Progenitor Stem Cell Research (ESPCR), Cairo, Egypt

Scientific illustrator: Abdullah Mahmoud

ELSEVIER

ACADEMIC PRESS
An imprint of Elsevier

Academic Press is an imprint of Elsevier
125 London Wall, London EC2Y 5AS, United Kingdom
525 B Street, Suite 1650, San Diego, CA 92101, United States
50 Hampshire Street, 5th Floor, Cambridge, MA 02139, United States
The Boulevard, Langford Lane, Kidlington, Oxford OX5 1GB, United Kingdom

ISBN: 978-0-12-821569-2

For Information on all Academic Press publications
visit our website at https://www.elsevier.com/books-and-journals

Figures were created using Biorender.com

Publisher: Stacy Masucci
Acquisitions Editor: Elizabeth Brown
Editorial Project Manager: Samantha Allard
Production Project Manager: Kiruthika Govindaraju
Cover Designer: Mark Rogers

Typeset by MPS Limited, Chennai, India

Dedication

This book is dedicated to our beloved family.

Contents

Part I Stem cells: basic biology 1

1. Introduction **3**

1.1 Establishing the concept of stem cells 3
Further reading 4

2. Anatomy and Histology **7**

2.1 Blood supply 8
2.2 Niche 10
2.3 Hemopoietic stem cell niches 11
2.4 Differences between osteoblastic and vascular niches 12
2.5 Components of the hemopoietic stem cell niche 12
2.6 Mesenchymal stem cell niche 14
2.6.1 Bone marrow mesenchymal stem cell niche 14
2.7 Mesenchymal stem niche in adipose tissue 15
2.7.1 Epidermal stem cell niche 16
2.7.2 Intestinal epithelial niche 16
2.8 Niche functions 17
2.9 Regulation of niche function 17
Further reading 19

3. Stem cells: definition, biological types, classifications, and properties **21**

3.1 Definition 21
3.2 Classification 21
3.3 Properties 22
3.3.1 Potency 22
3.3.2 Self-renewal 25
3.3.3 Stem cell plasticity 25
3.3.4 Evidence of stem cell transdifferentiation 27
3.3.5 Theories of transdifferentiation 27
3.4 Differentiation 29
3.5 Homing and engraftment 29
3.5.1 Definition 29
3.5.2 Mechanism of homing 30
3.5.3 Enhanced and directed homing 30

3.5.4 Stem cell trafficking 31
3.5.5 Mechanism of stem cell trafficking 31
3.5.6 Role of stem cell trafficking in regeneration 32
Further reading 32

Part II Stem cell induced regeneration: principles and applications

4. Mechanisms of stem cell–related regeneration 37

4.1 Regulation of tissue regeneration 38
4.2 Overview of the mechanism of regeneration 38
4.2.1 Transdifferentiation 39
4.2.2 Contribution to tissue regeneration 40
4.2.3 Variation in transdifferentiation potential 40
4.2.4 Molecular mechanisms of transdifferentiation 41
4.3 Signaling pathways involved in stem cell transdifferentiation 42
Mogrify algorithm 43
4.4 Activation of endogenous repair system 43
4.5 Induction of angiogenesis 44
4.5.1 Variability in angiogenic potential of mesenchymal stem cells 46
4.6 Secretory function 47
4.6.1 Microvesicles and exosomes 47
4.6.2 Paracrine action 48
4.7 Mitochondrial transfer 49
4.7.1 Inducers of transfer 50
4.7.2 Mechanisms 50
4.7.3 Effects 50
4.7.4 Variability in mitochondrial transfer according to mesenchymal stem cell source 50
4.8 Microenvironment modulation 51
4.9 Conclusion 53
Further reading 53

5. Sources of stem cells for regenerative purposes 57

5.1 Characteristics of the ideal source of stem cells 57
5.2 Sources of mesenchymal stem cells 58
5.3 Skin mesenchymal stem cells 62
5.4 Periodontal ligament stem cells 62

5.4.1 Sources of hemopoietic stem cells 62
5.5 Cord blood 63
5.5.1 Sources of endothelial cells 63
5.6 Tissue-specific stem cells 64
Further reading 66

6. Basic concepts in stem cell therapy: stem cell therapy strategy development 69

6.1 Introduction 69
6.2 What is the basic defect? 70
6.2.1 Cellular defects 70
6.2.2 Structural defects 71
6.3 Which cell lineage is the best for repair? Choosing optimum cell lineage 71
6.4 What is the best source for a cell type? Choosing the optimum cell source 74
6.4.1 Hemopoietic stem cells 74
6.4.2 Mesenchymal stem cells 74
6.5 What is the optimum laboratory protocol? 74
6.6 Choosing the optimum route of injection 76
6.7 Factors affecting homing of stem cells 77
6.7.1 Methods to improve stem cell homing 78
6.7.2 Mesenchymal stem cell homing 78
6.8 Optimal timing for cell therapy 79
6.9 Optimum number of cells for injection 79
6.10 Choosing the method of evaluation 80
Further reading 80

Part III Guidelines for stem cell therapy

7. Stem Cell Therapy: Laboratory guidelines 85

7.1 Introduction 85
7.1.1 Regulatory bodies 86
7.2 Laboratory guidelines 87
7.2.1 Structural facilities 88
7.2.2 Personnel qualifications and job description 94
7.3 Quality control team 96
Further reading 96

8. Protocol guidelines for clinical decisions in stem cell therapy 99

8.1 Introduction 99
8.1.1 Timing of intervention 99
8.1.2 Stem cell injection 99
8.1.3 Frequency of injection 101
Reference 103
Further reading 103

9. Protocol guidelines: laboratory decisions in stem cell therapy 105

9.1 Introduction 105
9.2 Choosing the cell purification level 106
9.3 Choice of purification strategy 106
9.4 Choice of stem cell differentiation status 107
9.5 Choice of cell dose 108
9.6 Choice of tracking method 109
9.6.1 Methods of in vivo cell tracking 109

10. Phases of translational stem cell research 111

10.1 Introduction 111
10.2 Laboratory experiments 111
10.3 Preclinical, experimental 112
10.3.1 Value of animal experiments 112
10.3.2 Small versus large animals 116
10.3.3 Challenges in animal experiments 117
10.3.4 Ethical considerations 117
10.4 Clinical trials 117
Further reading 118

11. Ethical guidelines of stem cell therapy 121

11.1 Introduction 121
11.1.1 Ethical guidelines for the use of embryonic cells 122
11.1.2 Ethical concerns 122
11.1.3 Ethical guidelines for preclinical studies 123
11.1.4 Principles of ethical guidelines of clinical translation 124
Further reading 125

12. Stem cell therapy: practical examples 127

12.1 Neuroregeneration 128

12.1.1 In vitro experiments 128
12.2 Experimental animal study 130
12.2.1 Induction of animal model 130
12.2.2 Stem cell therapy 131
12.2.3 Followup 131
12.3 Results 131
12.3.1 Clinical translational study 132
12.4 Results of clinical translational study 132
12.5 Replacement of defective protein 134
12.5.1 Epidermolysis bullosa 134
12.6 Results 135
12.7 Multifactorial defects 135
12.7.1 Results 136
12.8 Conclusion 137
Further reading 137

Appendix A *139*
Index *147*

PART I

Stem cells: basic biology

CHAPTER 1
Introduction

Content

1.1 Establishing the concept of stem cells 3
Further reading 4

1.1 Establishing the concept of stem cells

The concept of "stem cell" was introduced to explain the phenomenon of continuous replenishment of mature cell compartments by short-lived cells for example, skin cells, blood cells, and intestinal epithelium. This term was first used by the German biologist Ernest Haeckel in 1868 to describe two concepts: the evolution of multicellular organisms from a unicellular organism and the differentiation of various body tissues from one cell (fertilized ovum).

Until the 1980s, most of the research involving stem cells used the hemopoietic system as the prototype. The hemopoietic system is indeed an excellent example of the diverse, versatile, and hierarchal cell proliferation and differentiation system. In 1908, the Russian histologist Alexander Maksimov used the term stem cell in his model of hematopoiesis.

The term "stem cells" gradually gained grounds; in 1930 Sabin and coworkers published an article identifying stem cells in the bone marrow and provied that the radiation damages the chromatin of stem cells, which in turn causes radiation-induced anemia.

In 1957, Thomas and his team proved through their experiments on dogs that the bone marrow contains a cell type that can reconstitute hematopoiesis after irradiation. Although they did not use the term "stem cell," they proved its functional property.

"Stem cells" was earlier a concept with an abstract meaning. In 1963, Till and McCulloch's pioneering work established stem cells as a reality — something that can be seen, counted, and studied. In their experiment, they rescued lethally irradiated mice by escalating doses of bone marrow cells. Mice showed macroscopic splenic colonies, linearly correlating with

Stem Cell Therapy
DOI: https://doi.org/10.1016/B978-0-12-821569-2.00002-8

the number of cells infused. These colonies (called colony-forming unit-spleen CFU-S) comprised all lineages of hemopoietic cells.

Characterization of these cells through further experiments confirmed *clonality*, that is, all cells in the colony come from one cell. Serial transplantation of these cells proved their *self-renewal*.

A. Beginnings of bone marrow transplantation:

In 1968, Robert Good started bone marrow transplantation (BMT) as an application of all the accumulating research regarding hemopoietic stem cells (HSCs). This technique, first applied on congenital immune deficiency disorders, was met with much skepticism. Translation of the idea from animal models to human subjects faced many difficulties and occasional obstacles and failures. After five decades of thorough research, BMT has now become the standard treatment for many benign and malignant hematologic disorders, offered to up to 50,000 patients per year.

B. Nonhemopoietic effects of BMT

Bone marrow transplantation opened a number of new horizons in research through followup of transplanted patients, moving HSCs from hematologic repair cells to universal repair machine.

Followup of patients after BMT:

Followup of patients after BMT revealed the regeneration of damaged tissues outside the bone marrow. Patients with renal failure or liver cirrhosis showed improvement in their condition after HSC infusion. This finding suggested the extra-hematologic repair capacity of HSCs.

Autopsy of patients undergoing gender-mismatched BMT:

Patients who received gender-mismatched HSC infusion and became engrafted but died for other reasons were studied for cell tracking using FISH X/Y studies. Donor-derived mature cells were found in all body organs, implying that injected HSCs homed, engrafted, and matured in these organs. The percentage of donor-derived cells varied according to the physiologic state of the organ, being more in damaged organs.

These findings suggested that the infused HSCs functioned as a universal repair cell, and their concentration or homing to the bone marrow was because the damage in the bone marrow was more profound than that in other organs.

Further reading

de la Morena MT, Gatti RA. A history of bone marrow transplantation. Immunol Allergy Clin North Am 2010;30(1):1–15. Available from: https://doi.org/10.1016/j.iac.2009.11.005 PMID: 20113883.

Haeckel E. Naturliche Schopfungsgeschichte (Berlin: Georg Reimer). Orig Term "Stem Cell" Miguel Ramalho-Santos Holger Willenbring Cell Stem Cell 1868;1 July 2007.
Kumar R, Sharma A, Pattnaik AK, Varadwa PK. Stem cells: an overview with respect to cardiovascular and renal disease. J Nat Sci Biol Med 2010;1(1):43–52.
McCulloch EA, Till JE. Cytological demonstration of the clonal nature of spleen colonies derived from transplanted mouse marrow cells. Nature 1963;197(4866):452–4.
Sabin FR, Doan CA, Forkner CE. The production of osteogenic saromata and the effects on lymph nodes and bone marrow of intravenous injections of radium chloride and mesothorium in rabbits. J Exp Med 1931;56(2):267–89.
Thomas ED, Lochte HL, Lu WC, Ferrebee J. Intravenous infusion of bone marrow in patients receiving radiation and chemotherapy. N Engl J Med 1957;257:491–6.
Weissman IL. Clonal origins of the hematopoeitic system: the single most elegant experiment. J Immunol 2014;192(11):4943–4.

CHAPTER 2

Anatomy and Histology

Contents

2.1 Blood supply 8
2.2 Niche 10
2.3 Hemopoietic stem cell niches 11
2.4 Differences between osteoblastic and vascular niches 12
2.5 Components of the hemopoietic stem cell niche 12
2.6 Mesenchymal stem cell niche 14
  2.6.1 Bone marrow mesenchymal stem cell niche 14
2.7 Mesenchymal stem niche in adipose tissue 15
  2.7.1 Epidermal stem cell niche 16
  2.7.2 Intestinal epithelial niche 16
2.8 Niche functions 17
2.9 Regulation of niche function 17
Further reading 19

Stem cells are the physiological reservoir for normal cell turnover as well as tissue regeneration after injurious incidents. They represent the natural proliferative compartment for mature functioning cells. Stem cells are present in all body organs and tissues and are located in a special anatomical setup called the *niche*.

Hemopoietic stem cells (HSCs), the precursors of all blood lineages, namely, white cells, red cells, and platelets, are the most widely studied human stem cells. Due to their easy availability, they have been an easy target to study the biological properties of stem cells. Throughout the history of stem cell scientific research, studying HSCs has always been a step forward.

The concept of stem cells was first introduced to account for the continuous replenishment of rapidly proliferating cells such a blood cells.

In this chapter, while discussing biological properties of stem cells, we will discuss HSCs and mention differences in other lineages when they apply.

Stem Cell Therapy
DOI: https://doi.org/10.1016/B978-0-12-821569-2.00005-3

Bone marrow is composed of islands of hematopoietic tissue and adipose cells surrounded by blood sinusoids – all contained in a meshwork of trabecular bone.

Endosteal lining, in the form of flat bone-lining cells, covers both the inner surface of bone cavities as well as the outer surface of cancellous bone spicules. This endosteal lining is composed of osteoblasts and osteoclasts as well as a layer of reticular connective tissue (Figs. 2–1 and 2–2).

2.1 Blood supply

Blood supply of the hemopoietic organs (bone marrow, liver, and spleen) is unique in vascular anatomy and the pattern of blood flow. Anatomically, the vessels are *sinusoidal*—not capillary—resulting from anastomosis between capillaries and venous sinuses near the cortical bone. The nutrient artery enters the flat bone through the nutrient canals, splits into ascending and descending branches, and runs parallel to the long axis of the marrow cavity, giving rise to numerous thin-walled arterioles and capillaries. Approaching the cortical bone, arterioles open up and anastomose with venous sinuses. This results in a *circular pattern of blood flow* in the marrow cavity.

Bone marrow sinusoids are approximately 50-μ wide, with thin walls composed of a single layer of endothelial cells (ECs) without basement

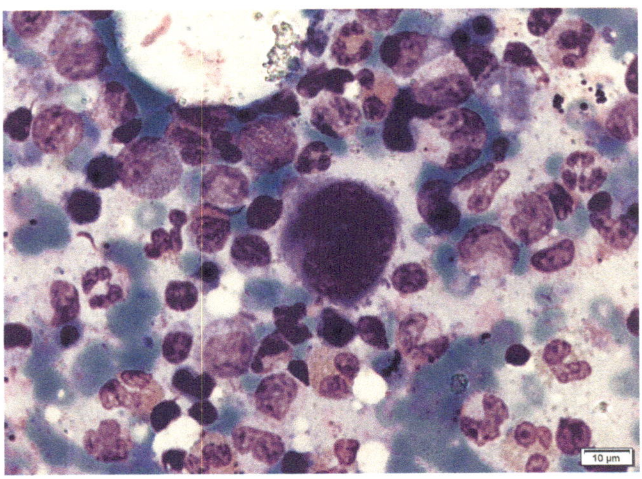

Figure 2–1 Bone marrow aspirate showing a collection of myeloid, erythroid, and megakaryocyte cells (Leishman–Giemsa stain, 100X).

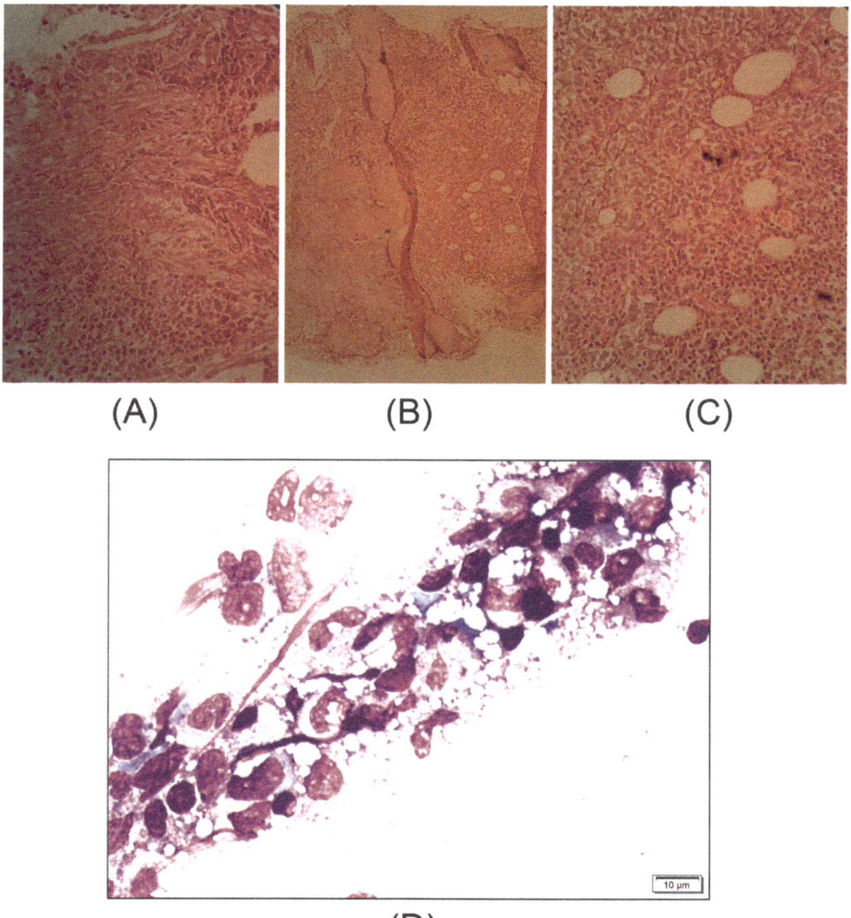

Figure 2–2 Bone marrow biopsy showing (A) compartments of cell collections (B) reticulin fibers (C) fat spaces (D) touch print smear.

membrane and arterioles open on most of their abluminal surface. Sinusoid walls allow trafficking of hematopoietic cells in both directions between blood and marrow. Sinusoids are formed by subdivision of blood-filled spaces of the marrow. Their size is larger than capillaries (20 μm) (Table 2.1).

There is no lymphatic drainage in bone marrow.

Bone marrow innervation: nerves enter the marrow space with arteries in the nutrient canal. Nerves follow arterioles, innervating smooth muscle of the vessels and terminating in the stroma forming part of the niche. The

Table 2.1 Bone marrow vasculature: A comparison between sinusoids and capillaries.

	Sinusoids	Capillaries
Size	50 μm	≤ 20 μm
Formation	Subdivision of blood-filled spaces	Ramification of arterioles
Wall	ECs with open pores and absent basement membrane	ECs and basement membrane
Cell trafficking	Allow passage of red cells, white cells, and plasma proteins	Allow only smaller molecules

significance of bone marrow innervation in stem cell anatomy lies in the link between overall body physiology and the stem cell "state," which can be quiescence, proliferation, trafficking, or differentiation.

Hematopoietic stroma, or microenvironment, is a network of cells, extracellular matrix (ECM) proteins, and fibers. Cells encompass stromal cells, such as adipocytes and reticular cells, bone lining cells, and macrophages.

Hematopoiesis is an anatomically regional process, both in relation to lineage as well as maturation. Each blood lineage has its unique anatomical location and anatomical subregion in the microenvironment.

2.2 Niche

Adult stem cells reside in a physiologically limited specialized anatomical microenvironment, or niche, which supports stem cells. Although the nature and location of the niche varies according to the tissue type, there are some common features of stem cell niche.

Common features of the niche:

1. Stem cell niche is a combination of cells, adhesion molecules, and ECM whose prime function is maintenance of the "stemness" state of the stem cell.
2. Stem cell niche is located in a special anatomical site in each organ, which is the optimum one for the biochemical needs of stem cells. This varies according to the function of each niche; in quiescent stem cell states, relative hypoxia is needed, and thus, the niche is located as far as possible from the blood supply. While in proliferating or committed stem cell states, niches are located adjacent to the blood supply.

In adult hematopoiesis, HSC "reservoir" in bone marrow is located in the osteoblastic niche on the bone surface. Upon mobilization, committed HSCs and their progenitor cells reside in the vascular niche adjacent to the sinusoid ECs. Thus, a gradient of the osteoblastic niche to the vascular niche exists in bone marrow, in which the osteoblastic niche provides a quiescent microenvironment, while the vascular niche favors commitment, proliferation, and differentiation (Table 2.2).

3. The niche functions as a "physical anchor" of stem cells through interaction between stem cells, niche cells, adhesion molecules, and ECM.
4. The niche controls stem cell fate determination, self-renewal, proliferation, and differentiation through secretion of extrinsic factors and signal molecules. This is done through an asymmetric distribution of extrinsic factors, signaling molecules, and their receptors. Therefore, stem cell proliferation depends on dynamic niche signaling. Maintaining a balance between the proliferation signal and antiproliferation signal is the key to homeostatic regulation of stem cells, allowing stem cells to undergo self-renewal while supporting ongoing tissue regeneration.

Thus, the niche is an anatomical entity, while the niche function is a physiological entity.

2.3 Hemopoietic stem cell niches

In mouse bone marrow, three anatomical regions (endosteal surfaces, sinusoidal walls, and hematopoietic tissue proper) and four cell lineages

Table 2.2 Comparison between osteoblastic and vascular BM-hemopoietic stem cell niche.

	Vascular niche	**Endosteal niche**
Site	Adjacent to Blood Vessels	Adjacent to the bone
Cellular components	Sinusoidal ECs	Osteoblasts
Adhesion molecules	E-selectin	Cadherins
Function	Self-renewal, proliferation, differentiation Migration and release	Maintenance of quiescence
Microenvironment	Richer in oxygen Richer in growth factors	Less oxygen concentration Less growth factors

(osteoblasts, sinusoidal ECs, adipocytes, and stromal cells) are involved. Bone marrow has a homogenous architecture with no structural and functional partitions. However, the bone marrow cavity is arranged in compartments having distinct properties. The three anatomical regions of bone marrow are the osteoblastic stem cell niche, the vascular stem cell niche, and the progenitor compartment.

HSCs are mainly maintained by the endosteal osteoblast niche, which provides a quiescent HSC microenvironment, while the vascular niche, can regulate the proliferation, differentiation, and mobilization of HSCs

The main task of osteoblastic niche is to maintain the long-term HSCs (LT-HSCs), which are capable of supporting hematopoiesis for months or even a lifetime as quiescent or low-cycling cells

2.4 Differences between osteoblastic and vascular niches

Researchers have proposed that a gradient of the osteoblastic niche to the vascular niche exists in bone marrow, with the osteoblastic niche providing a quiescent microenvironment, and the vascular niche favoring proliferation and further differentiation.

Characteristics of the endosteal region:

1. Comprises both bone-forming (osteoblasts) and bone-resorbing (osteoclasts) cells.
2. Enriched in microvessels, being the sites of distal ramification of blood vessels.
3. Site of anastomotic network between bone and marrow circulation, giving a specific ionic environment at the endosteum.
4. Enriched in stromal cells, which are also osteoprogenitors [most probably mesenchymal stem cells (MSCs)].

Characteristics of the vascular niche:

Vascular niche is the key player in hematopoiesis. It controls the interaction between ECs and HSCs, which regulates the balance between self-renewal and proliferation/differentiation. Committed HSCs and their progeny of differentiating progenitor cells leave the osteoblastic niche complete their differentiation in the vascular niche.

2.5 Components of the hemopoietic stem cell niche

HSC niche does not have a fixed anatomy; thus, it is continuously changing in adaptation with microenvironment. Interchange between the anatomical regions and the stromal as well as adhesions molecules (selectins, integrins, and laminin) forms the dynamic composition of the niche. The niche physiology is dependent on HSCs, stromal cells, ECM proteins, especially adhesion molecules, blood vessels, nerve supply, and ECs.

Components:

I. Cellular components:

1. Osteoblasts (osteoprogenitors):

Osteoblasts are bone-forming cells found in nests called osteons and derived from osteochondral progenitors, which are derived from the MSCs. Osteoblasts form an essential component of the stem cell niche.

2. ECs:

Bone marrow ECs are essential components of the stem cell niche. These cells express Notch ligands, which promote proliferation of early HSCs and prevent their exhaustion.

3. Adipocytes:

Despite the classical teaching that claims inhibitory role of adipocytes on hematopoiesis, recent research indicates that adipocytes promote hematopoietic recovery after bone marrow injury.

4. Pericytes (CD146 + MSC):

Pericytes are also known as adventitial reticular cells. They are multipotent cells with multipotent capacity of forming CFU-F, probably the mature counterpart of MSCs. Thus, they can differentiate into adipocytes, osteocytes, and skeletal muscle cells.

Pericytes are contractile cells that wrap microcirculation and venules controlling blood flow. They encompass a number of subtypes with variable expression of α-smooth muscle actin.

II. Adhesion molecules

Adhesion molecules are specialized proteins that maintain tissue integrity through cell—cell and cell—matrix adhesion. There are four major superfamilies of adhesion molecules: immunoglobulin superfamily, cadherins, integrins, and C-type lectin like domain proteins. Adhesion molecules play a crucial role in niche homeostasis. This is done through the following functions:

1. Stem cell retention:

Stem cell retention in the niche is mediated by cadherins, which are responsible for anchorage of stem cells to the niche. The cadherin expression level determines the availability of space in the niche.

2. Regulation of cell division:
   Self-renewal, a fundamental property of stem cells, enables the generation of differentiated cells without disturbance of the stem cell reserve. It is possible through asymmetric cell division. Cadherins and integrins regulate the plane of cell division in stem cells through interaction with the astral microtubules enabling centrosome positioning.
3. Regulation of stem cell trafficking:
   Stem cell trafficking occurs in two ways: (1) between niches, that is, moving from the quiescent endosteal niche to the dividing vascular niche; and (2) homing of stem cells to the niche. Both processes are regulated by adhesion molecules.

2.6 Mesenchymal stem cell niche

MSCs play an increasingly important role in stem cell therapy and regenerative medicine. Even though they are present in all body tissues, their source as a regenerative material remains confined to bone marrow, umbilical cord Wharton jelly, and adipose tissue.

2.6.1 Bone marrow mesenchymal stem cell niche

As with HSCs, there are two anatomical sites for MSC niches: periosteal and vascular (Table 2.3).

Table 2.3 Comparison between osteoblastic and vascular BM-mesenchymal stem cell niches.

	Osteoblastic	Vascular
Cells	Macrophages	ECs Neural cells
ECM	Elastin, collagen, periostin	Connexin
Signaling molecules	PDGF BMP FGF	PDFG-B, PDGF-D, BMP Notch pathway
Function	Promotes quiescence Proliferation in response to injury and stress (mechanosensing)	Promotes physiologic function of MSCs
Special mechanism	Mechanosensing	

A. Periosteal MSC niche:

The first documentation of this cell lineage was by Arnold Caplan in 1991 as a type of stem cell present in the periosteum with the ability to differentiate into osteogenic, chondrogenic, and adipogenic lineages. As in HSC niches, niche anatomy is comprised of cells, ECM fibers, and molecules. In addition to the biological and chemical composition, physical orientation of the structural collagen fibers of the periosteal niche play an important role in stem cell proliferation, especially after injury. Collagen fibers act as mechanosensors promoting proliferation during injury and stress, whereas the normal "steady-state" promotes stem cell quiescence. The physiology of the stem cell niche consists of the following:

a. Cells: macrophages, endothelial cells, fibroblasts

b. Fibers: elastin, collagen, periostin

c. Special mechanisms: mechanosensing

d. Signaling molecules; sclerostin (Sost), sclerostin domain-containing protein 1 (Sostdc1), platelet-derived growth factor (PDGF), bone morphogenic protein (BMP), fibroblast growth factor (FGF)

B. Vascular MSC niche:

MSCs associated with sinusoids and vessels are equivalent to "pericytes." Pericytes are multifunctional cells embedded in the basement membrane of the capillaries and venules in the body. They have been identified as in-vivo precursors of MSCs in many organs, with great debate regarding their functions during injury and homeostasis.

2.7 Mesenchymal stem niche in adipose tissue

Stem cell niche in the adipose tissue is composed of the same anatomical and cellular structure as in bone marrow. In addition, it is characterized by the presence of glycosaminoglycans (GAGs) in the ECM. GAGs' function is to protect stem cells from depletion and overproliferation. ECM in adipose stem cell (ASC) niche also contains type I, III, V, and VI collagen, adiponectin, fibronectin 1, and laminin. Growth factors influencing adipose stem cell niche function include vascular endothelial growth factor (VEGF) and PDGF.

MSC niches in other tissues are not as extensively studied, but they most likely share the same main characteristics as previously described niches.

2.7.1 Epidermal stem cell niche

Epidermal stem cell lineage is the precursor of skin appendages constituting the pilosebaceous unit: hair follicle (HF), sebaceous gland, and erector pili muscle. Each of these appendages have their own stem cell population. Thus, the pilosebaceous unit is comprised of heterogenous cell types and multiple stem cell pools, with probably multi- and unipotent progenitors.

HF as well as melanocyte stem cells are located in the bulge region of the HF, which is responsible for regeneration of the whole pilosebaceous unit. In addition to the primary function of niche cells, which is secretion of factors that either maintain quiescence or induce proliferation of stem cells, niche cells also sense environmental changes and adjust their activity to correlate with the body's variable physiological needs.

The bulge niche cellular component comprises adipocytes, MSCs, melanocyte stem cells, as well as blood vessels and nerves. Adipocytes are an important player in metabolic and nonmetabolic functions of the niche. Adipocytes produce antibacterial peptides during bacterial infection. An important function of adipocytes is sensing systemic environmental nutritional state and injuries. Adipocytes also induce HF stem cell proliferation and differentiation through secretion of PDGF.

MSCs are present as specialized groups in the HF bulge, known as dermal papilla (DP). DP MSCs release signaling molecules that control HF SC activity such as BMPs, FGFs, PDGFs, and Wnts.

2.7.2 Intestinal epithelial niche

Epithelial stem cells in the small intestine show unique compartments: stem cells and transient amplifying cells arranged in the crypts, while postmitotic and mature cells are arranged along the intestinal villi.

Cellular components of the intestinal stem cell (ISC) niche are divided into two groups: pivotal cells or adjacent cells. Adjacent cells are Paneth cells, smooth muscles, and neural and stromal cells. Pivotal niche cells are pericryptal myofibroblasts, fibroblasts, ECs, neural cells, pericytes, and smooth muscle cells. Paneth cells are the unique cells of the ISC niche. Paneth cells are multifunctional cells that secrete antibacterial peptides to sterilize the niche and express signaling molecules such as EGF, Notch ligand, and TGF-a.

The main components of the ISC niche are laminin, type IV collagen, and fibronectin, while the important signaling molecules are TGFB, BMP, Wnt, MMPs, bFGF.

2.8 Niche functions

Stem cell niche is a dynamic microenvironment with a specialized function. The main function of stem cell niche is to regulate the balance between the quiescent and proliferative states of stem cells by controlling self-renewal, proliferation, and differentiation. The stem cell niche also controls the choice of the "fate" of stem cell progeny. This is accomplished through an intricate interaction between stem cells, niche cells, ECM, and regulatory molecules in the microenvironment. Also, physical factors, such as temperature, oxygen tension, and mechanical factors are also involved. The communication between the niche and the overall body metabolic state is done through sensors in the niche.

2.9 Regulation of niche function

Niche function is regulated through niche cells, ECM, growth factors, cytokines, and biochemical conditions such as oxygen tension. Molecular pathways orchestrate the function of all these players. The main mechanisms involved in niche regulation are molecular pathways, growth factors, hormones, circadian rhythm, and oxygen level.

I. Molecular pathways:

1. CXCL12–CXCR4 signaling axis:

CXCL12, or stromal derived growth factor SDF-1, is a chemokine essential for HSC maintenance and retention in bone marrow. It is primarily expressed by perivascular mesenchymal stromal cells (CAR cells, nestin-green fluorescent protein (GFP) positive) and to a less extent by ECs and osteoblasts. Thus, the main activity of this axis is in the vascular niche. SDF-1 enhances transendothelial migration.

The function of SDF-1 varies according to the cells expressing it: its expression on the ECM induces transendothelial migration, and its expression on osteoblasts helps HSC homing to osteoblastic niche and quiescence, whereas its expression on ECs induces mobilization. Therefore, the role of SDF-1 in stem cell physiology is to retain stem cells in the bone marrow and maintain its quiescence and repopulating activity. SDF-1 is expressed on perivascular stromal cells, namely, SDF-1 abundant reticular cells, nestin-positive stromal cells, and leptin receptor–positive stromal cells.

2. Notch signaling:

Notch ligands are expressed both on ECs in the vascular stem cell niche and on HSCs. The upregulation of notch expression on ECs favors expansion of HSCs without differentiation.

II. Growth factors:

Growth factors secreted by ECs:

1. Stem cell factor:

Stem cell factor is secreted by ECs as well as bone marrow fibroblasts, osteoblasts, CXCL12-expressing perivascular stromal cells, and nestin-expressing MSCs.

2. Endothelial selectin:

E-selectin is expressed exclusively by bone marrow ECs in the vascular niche.

3. Glycoprotein 130:

Glycoprotein 130 (gp130), a signaling subunit shared by interleukin-6 (IL-6) family of cytokines receptors, has also been showed to make significant contribution to hematopoiesis.

III. Circadian rhythms:

Niche physiology shows strong relation with hormones affected by the circadian rhythm. Thus, HSC number, cell cycle status, as well as mobilization of bone marrow are affected by the circadian rhythm.

IV. Sympathetic nervous system signals

Sympathetic nervous system signals regulate bone formation through adrenergic signals.

V. Hormones and paracrine factors that regulate HSCs and the HSC niche:

Various body hormones contribute to the regulation of HSC physiology such as parathyroid hormones, estrogen, and prostaglandin E2.

VI. Oxygen gradient:

Oxygen concentrations play an important role in the bone marrow physiology. Normally, bone marrow microenvironment is a hypoxic microenvironment due to the low blood perfusion, allowing maintenance of higher concentrations of cytokines and growth factors to ensure their action on bone marrow matrix cells. Low oxygen concentrations favor quiescent cells with low metabolic requirements and limit reactive oxygen species production, which have a possible mutagenic effect on stem cells.

Further reading

Attwell D, Mishra A, Hall CN, O'Farrell FM, Dalkara T. What is a pericyte? J Cereb blood flow Metab: Off J Int Soc Cereb Blood Flow Metab 2016;36(2):451–5. Available from: https://doi.org/10.1177/0271678X15610340.

Caplan A. Mesenchymal stem cells. J Orthop Res 1991;9(5):641–50.

Chen C, Huang W, Wang E, Tai K, Lin S. Functional complexity of hair follicle stem cell niche and therapeutic targeting of niche dysfunction for hair regeneration. J Biomed Sci 2020;27 Article number: 43.

Ehninger A, Trumpp A. The bone marrow stem cell niche grows up: mesenchymal stem cells and macrophages move in. J Exp Med 2011;208(3):421–8.

Festa E, Fretz J, Berry R, Schmidt B, Rodeheffer M, Horowitz M, et al. Adipocyte lineage cells contribute to the skin stem cell niche to drive hair cycling. Cell 2011;146 (5):761–71.

Goldstein J, Horsley V. Home sweet home: skin stem cell niches. Cell Mol Life Sci 2012;69(15):2573–82.

Karina, Pawitan JA, Rosadi I. Adipose-derived stem cells and their microenvironment (Niche) in type 2 diabetes mellitus. Stem Cell Investig 2020;7:2020.

Kurenkova A, Medvedeva E, Newton P, Chagin A. Niches for skeletal stem cells of mesenchymal origin. Front Cell Dev Biol 2020;. Available from: https://doi.org/10.3389/fcell.2020.00592.

McCarthy N, Kraiczy J, Shivdasani R. Cellular and molecular architecture of the intestinal stem cell niche. Nat Cell Biol 2020;22:1033–41.

Meran L, Baulies A, Li V. Intestinal stem cell niche: the extracellular matrix and cellular components. Stem Cell Int 2017;2017 Article ID 7970385.

Moore S, Heu C, Yu N, Whan R, Knothe U, Milz s, et al. Translating periosteum's regenerative power: insights from quantitative analysis of tissue genesis with a periosteum substitute implant. Stem Cell Transl Med 2017;5:1749–59.

Nashchekina Y, Yudintceva N, Nikonov P, Ivanova E, Smagina L, Vorokina I. Effect of concentration of collagen gel on functional activity of bone marrow mesenchymal stromal cells. Bull Exp Biol Med 2017;163:123–8.

Ren G, Roberts AI, Shi Y. Adhesion molecules: key players in mesenchymal stem cell-mediated immunosuppression. Cell Adhes Migr 2011;5(1):20–2. Available from: https://doi.org/10.4161/cam.5.1.13491.

Tumbar T, Guiasch G, Greco V, Blanpain C, Lowry W, Rendl M, et al. Defining the epithelial stem cell niche in skin. Science 2004;281.

Wong V, Levi B, Rajadas J, Longaker M, Gurtner G. Stem cell niches for skin regeneration. Int J Biomater 2012;2012:8 Article ID 926059.

Yianni V, Sharpe P. Perivascular-derived mesenchymal stem cells. J Dent Res 2019;98 (10):1066–72.

CHAPTER 3

Stem cells: definition, biological types, classifications, and properties

Contents

3.1 Definition 21
3.2 Classification 21
3.3 Properties 22
3.3.1 Potency 22
3.3.2 Self-renewal 25
3.3.3 Stem cell plasticity 25
3.3.4 Evidence of stem cell transdifferentiation 27
3.3.5 Theories of transdifferentiation 27
3.4 Differentiation 29
3.5 Homing and engraftment 29
3.5.1 Definition 29
3.5.2 Mechanism of homing 30
3.5.3 Enhanced and directed homing 30
3.5.4 Stem cell trafficking 31
3.5.5 Mechanism of stem cell trafficking 31
3.5.6 Role of stem cell trafficking in regeneration 32
Further reading 32

3.1 Definition

Stem cells are defined as special human cells with the unique ability to differentiate into specialized cells while maintaining the original stem cell pool as a regenerative reserve.

3.2 Classification

Stem cells can be classified according to various approaches: the source of stem cell acquisition, stem cell's level in hierarchical steps or ontogeny,

Stem Cell Therapy
DOI: https://doi.org/10.1016/B978-0-12-821569-2.00003-X

types and biological properties, such as potency. The different types are listed below:

- **I.** According to the source of stem cell acquisition:
  - **A.** Cord blood and tissues
  - **B.** Bone marrow
  - **C.** Peripheral blood
  - **D.** Adipose tissue
  - **E.** Body fluids
  - **F.** Tissue stem cells
- **II.** According to ontogeny:
  - **A.** Embryonic
  - **B.** Fetal
  - **C.** Adult
  - **D.** Induced pluripotent
- **III.** According to biological properties:
  - **A.** Hemopoietic stem cells (HSCs)
  - **B.** Mesenchymal stem cells (MSCs)
  - **C.** Endothelial stem cells
  - **D.** Tissue-specific stem cells

3.3 Properties

3.3.1 Potency

Cell potency is defined as the ability of a living cell to differentiate into various cell types. Stem cells can express variable degrees of potency, according to which they can be classified as totipotent, pluripotent, multipotent, oligopotent, and unipotent cells. Cell potency is determined by the differential gene expression or activation of the cells. Totipotent stem cells are the cells with the highest differentiation potential, whereas pluripotent, multipotent, oligopotent, and unipotent express decreasing differentiation potential up to only one cell type in unipotent stem cells.

3.3.1.1 Totipotent stem cells

Totipotency is the ability of a single cell to differentiate into all adult somatic cell types as well as tissues of the placental and fetal membranes. The best example of totipotent stem cells is the zygote until the 16-cell stage, which is created after fertilization of the ovum. It then differentiates into any of the three germ layers: endoderm, ectoderm, and mesoderm in addition to placental cytotrophoblast and syncytiotrophoblast. After the

16-cell stage, approximately four days after fertilization, cells begin to lose their totipotency, becoming pluripotent embryonic stem cells.

Factors controlling totipotency: Mechanisms regulating totipotency are not clearly defined. However, possible mechanisms of gene activation are RNA regulation, interplay between miRNA and RNA binding proteins, or downregulation of chromatin assembly activity of CAF-1.

3.3.1.2 Pluripotent stem cells

Pluripotent stem cells are the ones that are capable of differentiation into all adult somatic cells, that is, all three germ layers: endoderm, mesoderm, or ectoderm. Pluripotent stem cells have two main determining features: the ability to form teratomas when injected in immune-deficient mouse and the ability to form chimera or contribution to the germline of mouse if injected into the blastocyst. Other features of pluripotency are characteristic colony morphology in culture, expression of pluripotency-associated transcription factors and reactivation of X chromosome.

Types of pluripotent stem cells:

1. Embryonic stem cells: A typical example of pluripotent stem cells is the embryonic stem cell arising from the inner cell mass of a fertilized ovum.
2. Artificially derived pluripotent stem cells: the first experiments reprogramming somatic cells into the pluripotent state used somatic cell nuclear transfer, introducing the nucleus of a differentiated cell into enucleated oocyte. Later, embryonic stem cells were used instead of oocytes. Subsequently, Yamanaka and Takahashi were able to reprogram adult cells by the introduction of four genes: Oct4, Saox2, Klf4, and cMyc, producing the induced pluripotent stem cell iPS.
3. Nonembryonic adult pluripotent stem cells: Certain studies support the presence of a rare subpopulation of pluripotent stem cells expressing SSEA-3 in bone marrow.

3.3.1.3 Molecular determinants of pluripotency

a. Genes

Master genes involved in the regulation and maintenance of pluripotency are Oct4, Sox2, and NANOG. Other genes such as KLF4, REX1, SSEA3, SSEA4, TRA-1−60, and TRA-1−81 coregulate pluripotency.

b. Epigenetic factors:

Pluripotency is induced and maintained by epigenetic factors inducing chromatin reorganization in an euchromatin state ensuring a more accessible state of the genes. This euchromatin state is found in the embryonic stem cells and induces pluripotent stem cells.

3.3.1.4 Cell cycle profile of pluripotency

Pluripotent stem cells show shortened G1 phase and a lack of G1 checkpoint regulation. Cyclins Ds are expressed at low levels in the pluripotent state. Normally, cyclins D1–3 expressed in G1 phase control the activity of CDK 4/6, which inhibits pRB and free E2F to initiate the G1–S transition.

3.3.1.5 Multipotent stem cells

Multipotency describes the ability of stem cells to differentiate into multiple but limited cell types. There are a number of examples of multipotent stem cells, such as HSCs, which can differentiate into lymphoid, myeloid, erythroid, and megakaryocyte precursors. MSCs are one of the most famous multipotent stem cells, and they are capable of differentiation into osteogenic, chondrogenic, and adipogenic cells. Thus, the most relevant cells for regenerative applications are multipotent stem cells.

3.3.1.6 Oligopotent stem cells

Oligopotent stem cells are able to differentiate into few cell types. Examples of oligopotent stem cells are the myeloid stem cells giving rise to neutrophils, eosinophils, and basophils, lymphoid stem cells giving rise to B and T cells, and vascular stem cells giving rise to endothelial and smooth muscle cells.

3.3.1.7 Bipotent stem cells

Further limitation of oligopotency is the bipotent stem cells giving rise to two types of cells. Hepatoblasts are the classic example of bipotent stem cells that give rise to hepatocytes and cholangiocytes.

3.3.1.8 Unipotent stem cells

Unipotent stem cells have the capacity to differentiate into only one cell type. Unipotent stem cells are rare and probably transient. Unipotent stem cells still retain the self-renewal capacity, which differentiates them from progenitor cells.

3.3.2 Self-renewal

Self-renewal, or the ability to maintain the reservoir of stem cells by undergoing specialized cell division giving rise to two different progenies (one identical stem cell and another destined to differentiation and proliferation), is the basic defining property of stem cells.

Two theories have been proposed to explain the mechanism of self-renewal:

1. Asymmetric cell division, in which differential segregation of fate determinants during cell cycle results in two different biological activities of the cells.
2. Stochastic differentiation, in which the stem cell divides into two identical stem cells; one undergoes differentiation, while the other remains quiescent, preserving the stem cell pool.

3.3.2.1 Regulation of self-renewal

"Stemness" refers to the two unique properties of stem cells: pluripotency and self-renewal. Regulation of stemness is a crucial process with implications on homeostasis, tissue regeneration, and carcinogenesis.

1. Intracellular regulation:

   Intracellular regulation occurs through the balance between three types of oncogenes: proto-oncogenes promoting self-renewal, gatekeeping tumor suppressors suppressing self-renewal, and care-taking tumor suppressors maintaining genomic integrity. The main intracellular regulator genes are NANOG, SOX2, Oct4, and KLF.
2. Extracellular regulation (niche regulation):

   Extracellular regulation of stem cell self-renewal is mediated through regulation of niche function. This is mainly undertaken by the cellular components of the niche: endothelial cells, pericytes, MSCs, as well as the extracellular matrix. Niche components control self-renewal by secretion of growth factors, especially leukemia inhibitory factor, cell—cell contact, cell—matrix adhesion, and miRNA.

3.3.3 Stem cell plasticity

Stem cell plasticity refers to the ability of stem cells to cross lineage boundaries to adopt the morphologic, antigenic, and functional characteristics of a different lineage outside their destined repertoire of differentiation (Fig. 3—1).

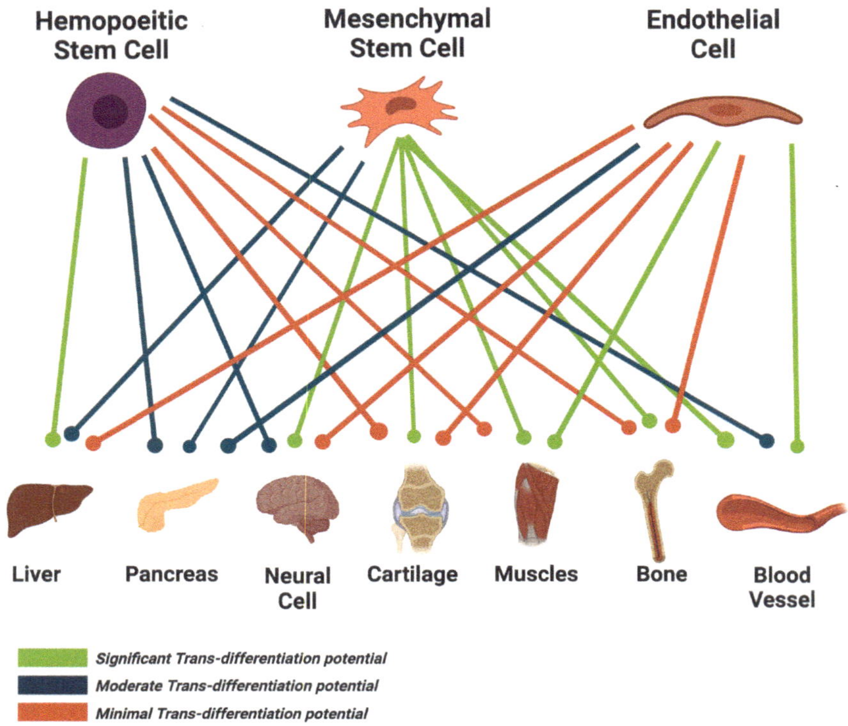

Figure 3–1 Examples of stem cell plasticity.

The difference between potency and plasticity:

Stem cell potency refers to the ability of stem cells to differentiate into a number of cell types according to its progeny, while plasticity refers to its ability to differentiate into cell types outside this normal progeny.

The difference between plasticity and transdifferentiation:

While plasticity is an inherent property of stem cells and plays an important role during embryogenesis, transdifferentiation refers to a comparable property during adult life. Transdifferentiation involves the conversion of cells from a specific tissue lineage into those from an entirely distinct lineage, with complete loss of the tissue-specific markers and functions of the original cell type and subsequent acquisition of markers and function of the new cell type. In other words, plasticity is the "embryonic" counterpart of transdifferentiation.

Early evidence of transdifferentiation came from followup of patients undergoing bone marrow transplantation. Regeneration of damaged organs outside the bone marrow was observed consistently. Autopsy studies from these patients revealed the presence of donor-derived mature cells in such organs. The uneven distribution of donor-derived cells between healthy and damaged organs led to the assumption that infused stem cells "homed" to these damaged tissues, not that their presence was the result of trafficking and entrapment. This was the first challenge to the concept of "lineage restriction" and the concept that HSC function is to replenish bone marrow and that their homing to bone marrow on the setting of bone marrow transplantation is due to the fact that it is the most injured organ in the body in this context.

These findings conformed with the general hypothesis of stem cell plasticity observed during embryogenesis.

3.3.4 Evidence of stem cell transdifferentiation

The term "transdifferentiation" was introduced after a series of observations following bone marrow transplantation. A large number of patients receiving bone marrow transplantation for a hematological disorder reported improvement in concomitant nonhematological diseases, such as renal impairment, liver failure, or cardiomyopathy. These observations triggered research on the mechanisms of homing and trafficking of HSCs.

Several studies have reported the presence of donor-derived mature cells in nonhematopoietic tissues such as skin, lung, liver, skeletal muscles, myocardium, kidney epithelium, and the central nervous system. HSC transdifferentiation into nonhematopoietic tissue was not restricted to systemic administration of HSC, after proof of such processes even after direct administration of HSCs into skeletal muscle.

3.3.5 Theories of transdifferentiation

The mechanisms of transdifferentiation may be one of the following:

1. True transdifferentiation:

   This means that genetic reprogramming of the cell occurred due to change in the microenvironment-based epigenetic factors resulting in differentiation into a completely different cell type expressing full morphologic, immunologic, and functional characteristics.
2. The presence of a "higher" stem cell with two possible fates: HSC and tissue-specific regenerative stem cell. This proposed cell should have the function of repair of any damaged cell type. It coexpressed HSC markers, and thus, it is isolated within the HSC in all experiments.

3. Cell fusion: Upon residence in a "nonhome" tissue, stem cells fuse with mature cells of the tissue resulting in a biological "cloning," with a progeny sharing the characteristics of stemness as well as tissue-specific markers.
4. Dedifferentiation: When stem cells are transplanted in a nonhome environment, they lose their genetic specificity and start differentiation along a new lineage.

3.3.5.1 Requirements for proof of transdifferentiation

As transdifferentiation has been considered an extraordinary event, extremely strict conditions are required to accept its occurrence. According to Wagers and Weissman (2004), the following criteria are mandatory to claim transdifferentiation:

1. Requirements in the cell of origin
   a. Definitive identification, quantification, and purification with extensive cell marker and genetic analysis.
   b. Minimal manipulation of the cells: using the cells directly after analysis without intervening culture.
2. Requirements in the final cell:
   a. Demonstration of concomitant expression of donor- and tissue-specific markers.
   b. Demonstration of integration into the target tissue.
   c. Demonstration of functional and immunological characteristics of the target tissue
   d. Demonstration of genetic reprogramming
3. Exclusion of cell fusion
4. Replication of the experiment by more than one laboratory using more than one experimental model.

3.3.5.2 Factors affecting frequency of transdifferentiation

There is a wide variation of the percentage of transdifferentiation in various reports, from less than 1% to about 20%. The most important determinants of the rate of transdifferentiation are the degree of tissue injury and host immunity.

3.3.5.3 Induction of transdifferentiation

In-vitro induction of transdifferentiation is an important cornerstone in the science of regenerative medicine and cellular therapy. Numerous approaches have been used in experimental induction of transdifferentiation, mimicking the physiologic maturation of specific cell lines. This can be done through stimulation of specific genes, using transcription factors, relevant cytokines, chemicals, simulation of specific microenvironment, or coculture with relevant cell lines.

3.3.5.4 Evaluation of transdifferentiation

To prove that transdifferentiation has occurred in an experimental procedure, the following must be proved:

1. Purification of the initial cell population to verify that the new cell lineage was not a component of the initial cell population.
2. Full characterization of the initial cell population: morphologic, immunologic, and genetic.
3. The resultant cells should show the following:
   a. Morphological features of the required tissue
   b. Immunological and surface markers of the required tissue
   c. Genetic profile (mRNA) of the required tissue
   d. Functional features of the required tissue

3.4 Differentiation

Stem cells can differentiate into functioning end cells. This is the normal "steady-state" physiologic function of stem cells throughout adult life. Stem cells replenish the cell population through replacement of damaged or dead cells. Stem cells proliferate and differentiate producing mature functioning cell population. This "maturation" is controlled by the homeostatic mechanisms of the body.

3.5 Homing and engraftment

3.5.1 Definition

Stem cell homing is the property by which circulating stem cells (endogenous or exogenous) can enter their corresponding environmental niche. During embryonic development, stem cell migration and homing is described in detail in organogenesis. In adult life, homing was described initially in HSCs and was applied in bone marrow transplantation procedure.

Homing is a crucial mechanism in normal hematopoiesis, bone marrow transplantation, as well as stem cell–based regenerative therapy

Stem cell homing is controlled by the SDF-1 and CXCR4. The SDF-1 and CXCR4 pathway regulation is crucial in the control of stem cell homing. Several factors have been involved in the regulation of these pathways, most of which are components of innate immunity and antimicrobial activity, which implicates a physiologic function of homing comparable to the neutrophil attractant systems. Examples of these factors are cleavage fragments of complement C3, prostaglandin E2, and antimicrobial peptides, such as B2-defensin, hyaluronic acid, thrombin and cathelicidin. These peptides enhance stem cell response to SDF-1.

3.5.2 Mechanism of homing

Homing is a multistep process starting by the interaction of stem cells with sinusoidal endothelium of the bone marrow. Endothelial cells slow cell movement in the sinusoids to allow cells to enter the marrow space through rolling via selectin and integrin receptors interaction with ligands on vascular endothelium such as vascular cell adhesion molecule (VCAM-1), intracellular adhesion molecule-1 (ICAM-1), fibronectin, and laminin. This initial interaction is weak and must be followed by CXCL12 and c-kit signaling inducing conformational changes in these integrins and upregulating the expression of VCAM-1 and ICAM-1. VLA-4/VCAM-1 interaction is the most important contributor to homing.

The next step is cell penetration of the sinusoidal wall either through the fenestrations between endothelial cells or through the endothelial cytoplasm. Stem cells can form podosomes to help transcellular migration. Stem cells then enter the bone marrow space and migrate between the adventitial cells to reach their final place "niche" (either osteoblastic or vascular).

CXCL12/CXCR4 axis is the principal pathway involved in homing of HSCs.

3.5.3 Enhanced and directed homing

As homing is the main determinant of effectiveness of stem cell therapy, enhancing and directing homing is needed. In case of HSC transplantation, homing is enhanced by increasing CXCR4 expression by coculture with stromal cells under hypoxia or modulation of CXCR4/CXCL12 axis. In MSC homing, which is also regulated by selectins and CXCL12, genetic overexpression of CXCR4 or VLA4 has been attempted. Also, direct administration of MSCs to target organs is possible. Other attempts to improve MSC homing include overexpression of chemokines in the target tissue.

3.5.4 Stem cell trafficking

Stem cell trafficking is an important process during embryogenesis with functions in organogenesis. This trafficking decreases as growth occurs, and it usually ends by the completion of embryogenesis.

In adult life, under normal physiologic states, stem cells reside in their niche and do not circulate in the bloodstream. HSCs do not exit bone marrow, and so do all other stem cells. However, a small percentage of HSCs are continuously entering the bloodstream, circulating through systemic circulation, and re-entering bone marrow. The role of this "physiologic mobilization and rehoming" is still unclear but is most probably linked to the regenerative and immunologic function of stem cells. This circulation has a circadian rhythm, which backs the idea that they function in homeostasis, patrolling the whole body for markers of tissue injury or inflammation.

A number of bioactive molecules govern this mobilization and trafficking, including chemokines such as SDF-1, growth factors, and bioactive lipids. There is accumulating evidence that these factors govern trafficking of non-HSCs such as MSCs, EPCs (Endothelial progenitor cells), and very small embryonic like cells (VSELs).

Complement proteins are an important regulator of stem cell trafficking, and complement cascade is an important sensor of homeostasis.

3.5.5 Mechanism of stem cell trafficking

Stem cell trafficking occurs in the following steps:

1. Detachment of stem cells from the niche: Stem cells are anchored in the niche through an interaction between SDF-1 and VCAM-1 in the niche matrix with CXCR4 and $\alpha4\beta1$integrin (VLA-4) in stem cells. Detachment is initiated by proteolysis of this interaction through complement cascade protein (C5a), which activates the release of proteolytic enzymes from neutrophils and monocytes.
2. Stem cells released from the niche are attracted through chemoattractants in the bloodstream. Thus, SDF-1gradients initiate stem cell egress into the blood, crossing the marrow–blood barrier.

Negative regulation of trafficking: Physiological retention of stem cells in bone marrow is mediated through heme oxygenase-1, which is an antiinflammatory enzyme.

3.5.6 Role of stem cell trafficking in regeneration

Circulating stem cells play an important role in tissue regeneration, a process that is utilized and enforced in stem cell therapeutics. In a splendid study of prognostic factors in myocardial infarction, Patel and his colleagues noted a variable increase in the number of circulating stem cells in patients with myocardial infarction as compared with normal controls. Followup of these patients proved a positive correlation between the number of circulating stem cells (increased physiologic endogenous stem cell mobilization) and favorable outcome in these patients (healing or regeneration).

Further reading

Baron MH. Embryonic origins of mammalian hematopoiesis. Exp Hematol 2003;31 (12):1160–9.

Biehl JK, Russell B. Introduction to stem cell therapy. J Cardiovasc Nurs 2009;24 (2):98–105. Available from: https://doi.org/10.1097/JCN0.0b013e318197a6a5.

Chute JP. Stem cell homing. Curr Opin Hematol 2006;13(6):399–406. Available from: https://doi.org/10.1097/01.moh.0000245698.62511.3d PMID: 17053451.

He S, Nakada D, Morrison SJ. Mechanisms of stem cell self-renewal. Annu Rev Cell Dev Biol 2009;25:377–406. Available from: https://doi.org/10.1146/annurev.cellbio.042308.113248.

Hima Bindu A, Srilatha B. Potency of various types of stem cells and their transplantation. J Stem Cell Res Ther 2011;1:115. Available from: https://doi.org/10.4172/2157-7633.1000115.

Ishiuchi T, Enriquez-Gasca R, Mizutani E, Boskovic A, Ziegler-Birling C, Rodriguez-Terrones D, et al. Early embryonic-like cells are induced by downregulating replication dependent chromatin assembly. Nat Struct Mol Biol 2015;22(9):662–71. Available from: https://doi.org/10.1038/nsmb.3066 PMID 26237512.

Komarova NL. Principles of regulation of self-renewing cell lineages. PLoS One 2013;8 (9):e72847.

Kristocheck M, Dias LD, Ghem C, et al. The homing of bone marrow stem cells is differentially activated in ischemic and valvular heart diseases and influenced by beta-blockers. J Transl Med 2018;16:133. Available from: https://doi.org/10.1186/s12967-018-1520-9.

Liesveld J, Sharma N, Aljitawi O. Stem cell homing: from physiology to therapeutics. Stem Cell 2020;38:1241–53.

Mitalipov S, Wolf D. Totipotency, pluripotency and nuclear reprogramming. Adv Biochem Eng/Biotechnol 2009;114:185–99. Available from: https://doi.org/10.1007/10_2008_45. Available from: doi: Bibcode:2009esc.book.185M. ISBN 978-3-540-88805-5. PMC 2752493. PMID 19343304.

Morgani SM, Canham MA, Nichols J, Sharov AA, Migueles RP, Ko MS, et al. Totipotent embryonic stem cells arise in ground-state culture conditions. Cell Rep

2013;3(6):1945–57. Available from: https://doi.org/10.1016/j.celrep.2013.04.034 PMC 3701323. PMID 23746443.

Morrison S, Shah N, Anderson D. Regulatory mechanisms in stem cell biology. Cell 1997;8(3):287–98.

Nagasawa T, Hirota S, Tachibana K, et al. Defects of B-cell lymphopoiesis and bone-marrow myelopoiesis in mice lacking the CXC chemokine PBSF/SDF-1. Nature 1996;382(6592):635–8.

Ng TK, Pelaez D, Fortino VR, Greenberg J, Cheung HS. Pluripotent adult stem cells: a potential revolution in regenerative medicine and tissue engineering. In: Bhartiya D, Lenka N, editors. Pluripotent stem cells [Internet]. London: IntechOpen; 2013 [cited 2022 Sep 16]. Available from: https://www.intechopen.com/chapters/45129. https://doi.org/10.5772/54366.

Niwa Hitoshi. Mechanisms of stem cell self-renewal. Chapter 7. In: Lanza Robert, Atala Antony, editors. Essentials of stem cell biology. Academic Press; 2014. p. 81–94.

Park C, Lewis A, Chen T, Lacorazza D. Regulation of self-renewal in normal and malignant hematopoeitic stem cells by Kruppel-like factor 4. Stem Cell Transl Med 2019;8 (6) 568–574.

Patel R, Li Q, Ghasemzadeh N, Eapen D, Moss L, Janjua A, et al. Circulating $CD34^+$ progenitor cells and risk of mortality in a population with coronary artery disease. Circul Res 2015;116:289–97.

Pauklin S, Vallier L. The cell-cycle state of stem cells determines cell fate propensity. Cell 2013;155(1):135–47. Available from: https://doi.org/10.1016/j.cell.2013.08.031 PMCID: PMC3898746 PMID: 24074866.

Polejaeva I, Mitalipov S. Stem cell potency and the ability to contribute to chimeric organisms. Reproduction 2013;145(3):R81–8. Available from: https://doi.org/10.1530/REP-12-0396 PMCID: PMC3678546 NIHMSID: NIHMS475822 PMID: 23221011.

Ratajczak M. A novel view of the adult bone marrow stem cell hierarchy and stem cell trafficking. Leukemia 2015;29:776–82. Available from: https://doi.org/10.1038/leu.2014.346.

Ratajczak MZ, Abdelbaset-Ismail A. Stem cell homing. In situ tissue regeneration. Academic Press; 2016. p. 21–34.

Ratajczak MZ. Stem cells and mechanisms regulating their trafficking – a new and challenging area of investigation in modern psychiatry. Prog Neuro-Psychopharmacol Biol Psychiatr 2018;80(Pt A):1–2. Available from: https://doi.org/10.1016/j.pnpbp.2017.08.018.

Romito A, Cobellis G. Pluripotent stem cells: current understanding and future directions. Stem Cell Int 2016;2016:20 Article ID 9451492.

Seydoux G, Braun RE. Pathway to totipotency: lessons from germ cells. Cell 2006;127 (5):891–904. Available from: https://doi.org/10.1016/j.cell.2006.11.016 PMID 17129777.

Seymour T, Twigger A-J, Kakulas F. Pluripotency genes and their functins in the normal and aberrant breast and brain. Int J Mol Sci 2015;16(11):27288–301 PMCID: PMC4661882 PMID: 26580604.

Sohni A, Verfaillie C. Mesenchymal stem cells migration, homing and tracking. Stem Cell Int 2013; Article ID 130763.

Takahashi K, Yamanaka S. Induction of pluripotent stem cells from mouse embryonic and adult fibroblast cultures by defined factors. Cell 2006;126(4):663–76.

Wagers AJ, Weissman IL. Plasticity of adult stem cells. Cell 2004;116(5):639–48. Available from: https://doi.org/10.1016/s0092-8674(04)00208-9 PMID: 15006347.

PART II

Stem cell induced regeneration: principles and applications

CHAPTER 4

Mechanisms of stem cell—related regeneration

Content

4.1 Regulation of tissue regeneration 38
4.2 Overview of the mechanism of regeneration 38
  4.2.1 Transdifferentiation 39
  4.2.2 Contribution to tissue regeneration 40
  4.2.3 Variation in transdifferentiation potential 40
  4.2.4 Molecular mechanisms of transdifferentiation 41
4.3 Signaling pathways involved in stem cell transdifferentiation 42
  Mogrify algorithm 43
4.4 Activation of endogenous repair system 43
4.5 Induction of angiogenesis 44
  4.5.1 Variability in angiogenic potential of mesenchymal stem cells 46
4.6 Secretory function 47
  4.6.1 Microvesicles and exosomes 47
  4.6.2 Paracrine action 48
4.7 Mitochondrial transfer 49
  4.7.1 Inducers of transfer 50
  4.7.2 Mechanisms 50
  4.7.3 Effects 50
  4.7.4 Variability in mitochondrial transfer according to mesenchymal stem cell source 50
4.8 Microenvironment modulation 51
4.9 Conclusion 53
Further reading 53

Regeneration refers to the renewal of a particular tissue or organ. The human body has natural reparative and homeostatic mechanisms, which deal with tissue injury signals. The philosophy of regenerative medicine is to augment and harness these natural repair mechanisms.

Stem Cell Therapy
DOI: https://doi.org/10.1016/B978-0-12-821569-2.00013-2

4.1 Regulation of tissue regeneration

Regenerative capacity varies according to various physiological and environmental factors, including age, growth pattern, body size, and life stage. Regenerative capacity decreases by aging due to cellular senescence, impaired cell proliferation, and cycling, as well as telomere shortening. Environmental factors include the degree of innervation, blood supply, as well as the type of tissue matrix.

4.2 Overview of the mechanism of regeneration

According to their regenerative capacity, human tissues vary from almost no regeneration (central nervous system and heart) to a slow cell turnover (liver and pancreas) to a rapid cell turnover (intestinal epithelium and skin) (Fig. 4–1).

Research involving stem cell–induced tissue regeneration works in many directions, questioning the varying regenerative potential of different stem cell lineages and sources as well as investigating the mechanisms of this regeneration in an attempt to control and augment this function (Fig. 4–2).

Since the beginning of the regenerative medicine era, and throughout most of the 1990s and the first decade of the 21st century, experts have strived to prove the transdifferentiation potential of stem cells, in both in vitro and animal experiments. Exciting positive results were found with almost all tackled organs. With advancements in preclinical and clinical

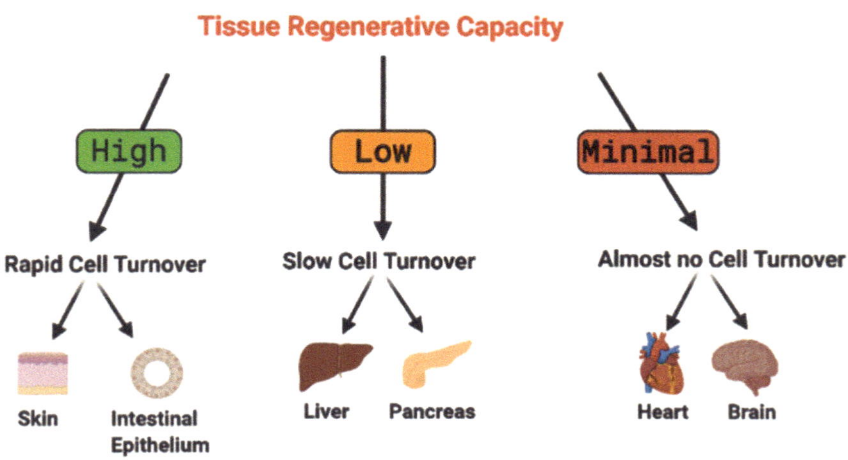

Figure 4–1 Classification of tissues according to their regenerative capacity.

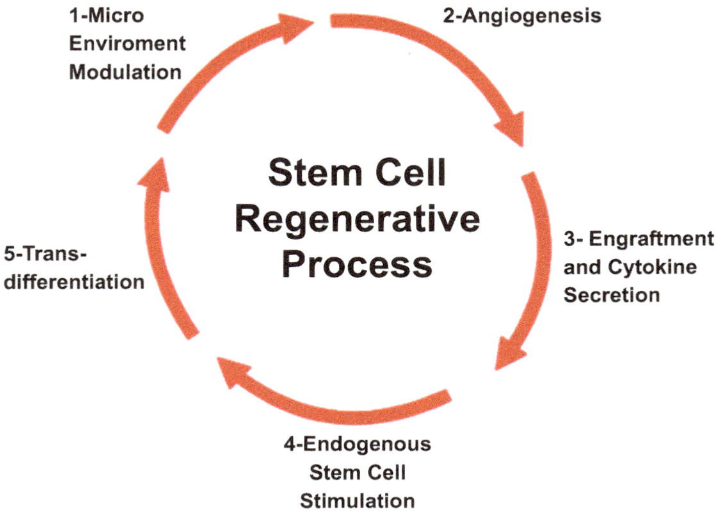

Figure 4–2 Steps of the stem cell Regenerative process.

models, the discordance between the rate of transdifferentiation and the degree of clinical improvement pointed to the wider mechanisms of tissue regeneration.

Early studies involving stem cell therapy in animal models of various degenerative disorders documented the positive clinical improvement following stem cell injection. Because these studies were based on in vitro experiments verifying the multipotent potential of stem cells, direct transdifferentiation of the injected stem cells into the defective cell lineage was suggested. However, later studies observed the lack of correlation between the extent of functional improvement and the actual cellular engraftment or transdifferentiation.

Cellular mechanisms contributing to stem cell–induced regeneration include transdifferentiation, induction of angiogenesis, activation of tissue-resident stem cells, paracrine functions, cell fusion, immunomodulation, regulation of inflammation, transfer of mitochondria, inhibition of fibrosis, and microenvironment modulation. The relative importance of each of these mechanisms depends on the tissue type, the extent and type of injury, as well as the type and dose of stem cells (Fig. 4–3).

4.2.1 Transdifferentiation

The phenomenon of transdifferentiation, the ability of stem cell to adopt antigenic, morphologic, and functional characteristics of a new lineage, is

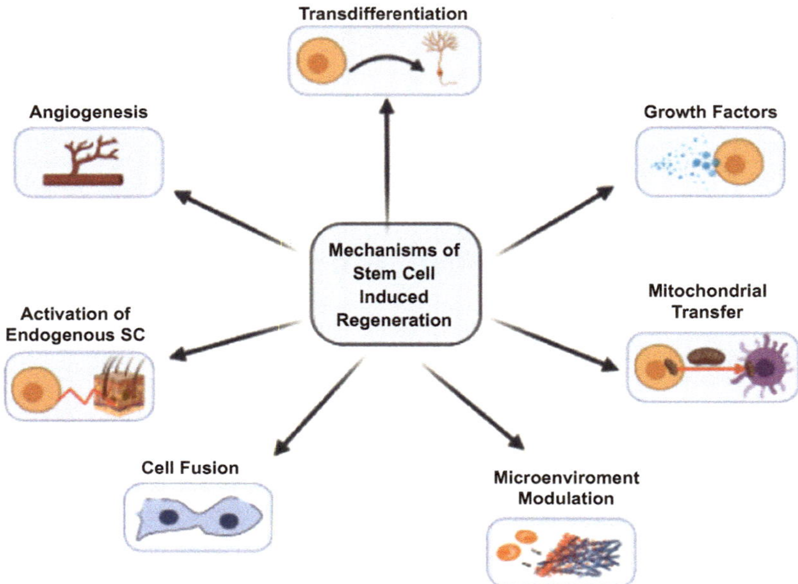

Figure 4–3 Mechanisms of stem cell–induced regeneration.

fascinating. The ability of the cells to cross lineage boundaries and break all the rules of traditional biology ignited the imagination of biologists. Early research used embryonic stem cells, while later research documented this property in almost all adult stem cells and even dedifferentiated adult mature cells.

Several reports have documented the in vitro transdifferentiation of both embryonic and mesenchymal stem cells (MSCs) into neural cells, cardiac cells, hepatic cells, pancreatic islet cells, retinal epithelial cells, and almost all body tissues.

4.2.2 Contribution to tissue regeneration

Despite the definitive occurrence of transdifferentiation, its contribution to stem cell–related regeneration is disputed. This is due to two main reasons: the minute percentage of documented transdifferentiation and the lack of correlation between clinical improvement and the number of transdifferentiated cells.

4.2.3 Variation in transdifferentiation potential

Although plasticity is one of the inherent biological properties of all stem cells, lineage and source variability infer different transdifferentiation patterns. Regarding lineage variability, hemopoietic stem cells (HSCs) show more potent hepatic differentiation, while MSCs have more neural transdifferentiation potential (Fig. 4–4).

4.2.4 Molecular mechanisms of transdifferentiation

Transdifferentiation, like metaplasia, is associated with defined changes in cellular morphology accompanied by changes in pattern of gene expression. The molecular mechanisms of transdifferentiation comprise two main factors, namely, genetic switching of the master regulatory gene to conform with that of the new cell type and epigenetic modification in favor of the pattern of the new lineage.

1 Master switch genes:

A master switch gene is the gene whose activation is both necessary and sufficient to activate a combination of genes in a coordinated manner leading to the development of a certain cell lineage. A classic example of master switch gene is MyoD, the master switch gene for myogenesis. For transdifferentiation within the same germ layer, master gene switching is a one-step process, while for different germ layers, it is a two-step process.

2 Epigenetic modification:

Epigenetic modifications through chromatin remodeling, histone modification, DNA methylation, and noncoding RNA

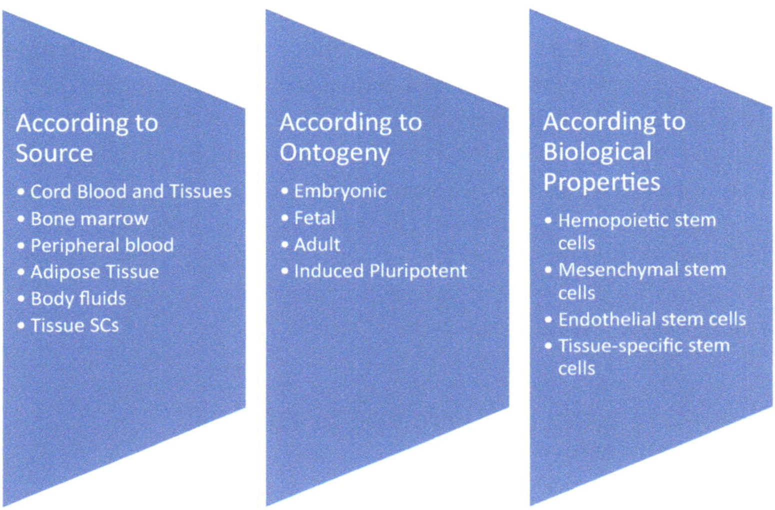

Figure 4–4 Approaches to classification of stem cells.

regulation are key players in determining stem cell fate. Transdifferentiation is implemented through reprogramming of the cells epigenetic print to establish new lineage-specific modifications. The following forms of chromatin remodeling play important roles in lineage re-determination:

a. ***Histone modification:*** Postranslational histone modifications include acetylation, methylation, phosphorylation, and ubiquitylation, the most common of which is acetylation. Histone acetylation is related to gene activation, whereas hypoacetylated genes are inactive. Histone acetylation weakens the interaction between histone tails and DNA, creating a space for factors to bind to the promotor initiating gene transcription. Thus, global histone acetylation is essential for transdifferentiation.

   Histone methylation is another modification responsible for shaping the map of active and silent genes.

b. ***DNA methylation:*** DNA methylation is an important epigenetic mechanism essential for normal cell development. It is fundamental for gene imprinting, X chromosome inactivation, and lineage-specific gene expression. DNA methylation occurs by adding a methyl group to cytosine within a CpG dinucleotide through DNA methyltransferases. The status of CpG methylation in the cell genome reflects their transdifferentiation potential, exposing the lineage. Thus, hyperacetylated and hypomethylated genes respond easily to lineage-specific inducers in the microenvironment.

c. ***Noncoding RNA regulation:*** During normal cell differentiation, multiple genes are expressed sequentially at each stage. Feedback and feedforward pathways regulate these stages tightly. MicroRNAs (miRNAs) are the mediators of this regulation. Specific miRNAs were reported in adipose-derived MSC transdifferentiation into the neural lineage, such as miR-22, miR-145, and miR-224.

Collectively, hyperacetylation and hypomethylation of the DNA prepare the genes for change in microenvironment stimuli, whereas miRNAs guide the lineage commitment steps.

4.3 Signaling pathways involved in stem cell transdifferentiation

Similar to natural development, signaling pathways active in transdifferentiation differ from one lineage to the other. Despite this lineage variation, a set of signaling pathways are implicated in stem cell fate determination. These are Wnt and B–Catenin pathway, Notch pathway, transforming growth factor beta (TGFB) pathway, bone morphogenetic protein (BMP) pathway, and sonic hedgehog pathway. While Notch signaling maintains stem cell self-renewal and inhibits differentiation and must be downregulated during transdifferentiation, Wnt and B–Catenin pathway, TGFB pathway, BMP pathway, and sonic hedgehog pathway are upregulated during transdifferentiation.

As a practical example of signaling pathways active during transdifferentiation, Ye et al. (2015) showed that MSC transdifferentiation into hepatic lineage passes through the following steps:

1. Change from mesoderm to endoderm: nodal signaling and canonical Wnt pathways.
2. Specification of hepatoblast: Activin/nodal signaling enhance endoderm determination followed by activation of Wnt/B–catenin signaling.
3. Determination of hepatocytes: TGFB and WNT signaling.

Mogrify algorithm

Cellular reprogramming experiments need consecutive combinations of growth factors to guide the cells through these lineage conversion steps. Mogrify algorithm is a network-based designed to find the set of transcription factors needed for conversion of one cell lineage to the other.

4.4 Activation of endogenous repair system

Injected stem cells can augment the resident stem cell–related repair mechanisms. These include a number of features favoring augmentation of the resident stem cell pool (Fig. 4–5).

A. Endogenous stem cell activation:

Using genetic lineage mapping, Loffredo et al. (2011) documented that injected stem cells increased the pool of endogenous cardiac progenitors.

B. Increasing the rate of proliferation of differentiated cells:

Hatzistgeros et al. (2010) proved the effect of stem cells on increasing the cycling rate of both undifferentiated and differentiated cardiac cells, leading to increased mature myocytes

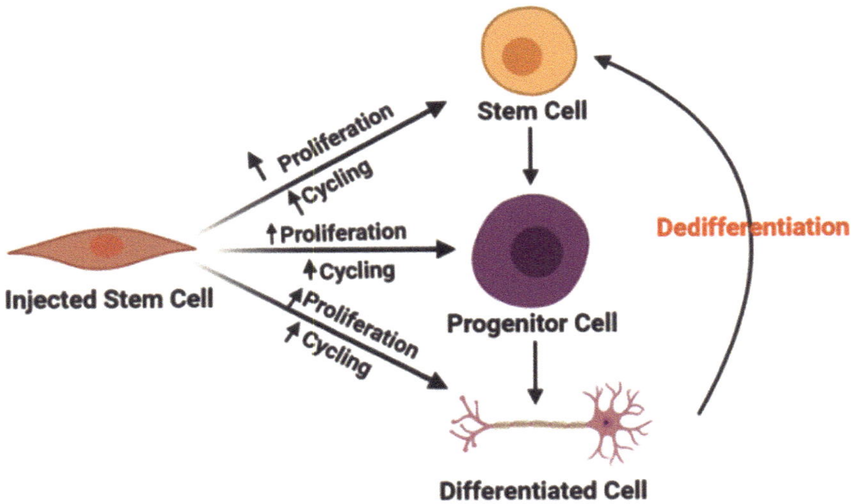

Figure 4–5 Effect of stem cells on endogenous cell populations.

C. Induction of dedifferentiation:

Dedifferentiation maybe an initial step in stimulation of endogenous repair by increasing the number of stem cells by recruitment of differentiated cells to enter the stem cell pool. This increases the rate of proliferation of the required cell lineage.

4.5 Induction of angiogenesis

MSCs are localized in the vascular niche in bone marrow, as well as around adult vessels, where they are called pericytes. MSCs play a pivotal role in regulating new blood vessel formation through a number of mechanisms.

Angiogenesis, or neovascularization, is a key mechanism of tissue regeneration. As a reaction to tissue damage or inflammation, inflammatory cytokines induce the evolution of new vascular networks to enhance the incoming "defense" cells and hasten the removal of toxic byproducts. Thus, harnessing this property is one of the stem cell mechanisms of endorsing endogenous repair mechanisms. This is mediated through one of the following:

A. Neovascularization through direct transdifferentiation into vascular lineage;

B. Secretion of "pro-angiogenic factors": IL-8, insulin-like growth factor I (IGF-1) and vascular endothelial growth factor (VEGF);
C. Cell–cell contact with endothelial lineage;
D. Paracrine-mediated activation of the angiogenic function of endothelial cells (ECs) through induction of VEGF-A signaling pathway (Table 4.1)
E. Exosome-mediated angiogenesis through miRNA.

Table 4.1 Paracrine angiogenic factors secreted by mesenchymal stem cells and proposed mechanism.

Factor	Mechanism
bFGF	organization of various angiogenic pathways and coordination of cell–cell interactions in this process
VEGF	Endothelialization of injured tissue by: 1 Inducing endothelial cell proliferation 2 Regulating ECs migrations 3 Promoting EC differentiation
TGF-β	Endothelial differentiation of MSCs
PDGF	Induces endothelial differentiation of MSCs:
Angiopoietin1	1 vascular protective effects 2 suppressing plasma leakage 3 inhibiting vascular inflammation 4 preventing endothelial death
Placental growth factor	Synergistic function with VEGF
IL-6	1 Potent antiapoptotic activity 2 Induction of VEGF expression on EC chemoattractant for angiogenesis
Monocyte chemotactic protein-1 (MCP-1)	
Stromal-derived factor (SDF-1)	Stimulates EC proliferation and capillary tube formation
Hepatocyte growth factor (HGF)	Induces activation of angiogenesis through tyrosine phosphorylation c-Met receptor found in ECs and SMCs
Artemin	Promotes angiogenesis
CXCL16	Promotes angiogenesis
Endothelin-1	Pro-angiogenic and pro-lymphangiogenic
IL-1B	Pro-angiogenic and porlymphangiogenic
IL-8	Pro-angiogenic

bFGF, basic fibroblast growth factor; *ECs*, endothelial cells; *MSCs*, mesenchymal stem cells; *PDGF*, platelet-derived growth factor; *SMCs*, smooth muscle cells; *TGF-β*, transforming growth factor beta; *VEGF*, vascular endothelial growth factor.

Cell-free MSC-derived exosomes induce enhancement of endothelial functions: proliferation, migration, and tube formation. This is mediated through pro-angiogenic micro RNAs: miRNA-125a, miRNA-30b, miRNA-30c, miRNA-424, miRNA-150, and let-7f.

On the other hand, MSC-derived secretome contains some antiangiogenic factors such as thrombospondin-1 and 2, serpin B5, E1 and F1, platelet factor 4, and vasohibin. Factors affecting the relative abundance of pro- and antiangiogenic factor are still under research.

4.5.1 Variability in angiogenic potential of mesenchymal stem cells

While some studies have reported antiangiogenic properties of MSCs, others have reported variable angiogenic/antiangiogenic properties across different MSC sources, for example, bone marrow and cord blood. This variability can be explained by a number of causes:

1. Source of MSCs:

   Angiogenic potential of MSCs is dependent upon their tissue of origin. MSCs from adipose show significantly more efficient angiogenesis potential than cord blood–derived MSCs. Amniotic fluid–derived MSCs are more potent angiogenic inducers than bone marrow–derived MSCs. This variability is due to increased amount of angiogenic factors in their secretome caused by microenvironment differences.
2. Culture conditions of MSCs: Exposure of MSCs to variable environmental conditions during culture induces modulation of key signaling pathways in MSCs. An excellent example of the effect of environmental factors on the regenerative and angiogenic properties of MSCs is the case of intrauterine growth restriction, where abnormal placental development induces placental MSCs to lose angiogenic potential while enhancing their adipogenic potential due to the metabolic shift from aerobic to anaerobic state. Also, Rezaie and coworkers reported reduction of the angiogenic potential of MSCs after exposure to diabetic sera.
3. Since the exosome-mediated pro-angiogenic secretome of the MSCs increases with hypoxia, maybe the mode rather than the actual effect of MSCs changes with the microenvironment. That is to say that in a hypoxic microenvironment, there is a shift of the pro-angiogenic factors of MSCs from paracrine to exosome mediated, which is a more intimate cell-to-cell contact and is protected from hostile milieu.

4.6 Secretory function

Most of the regenerative effects of stem cells are mediated through their library of growth factors, cytokines, chemokines, bioactive lipids, messenger RNA (mRNA), and regulatory miRNA. This secretory function is crucial for autocrine and paracrine regulation of cellular functions. Stem cells activated by mobilization from their physiological niches secrete these regulatory molecules. These bioactive molecules are either secreted directly into the microenvironment or "shuttled to the cells through microvesicles and exosomes" (Table 4.2).

4.6.1 Microvesicles and exosomes

Microvesicles are shed from the cell surface as a physiological phenomenon accompanying cell activation or exposure to injuries such as hypoxia, irradiation, complement components, and oxidative injury. The size of microvesicles ranges from 0.1 to 1 μm.

In contrast, exosomes are formed in the endosomal membrane and released as exocytosis. The size of exosomes is much smaller than MV, ranging from 30 to 100 nm.

Table 4.2 Comparison between main characteristics of microvesicles and exosomes.

	MV	Exosomes
1. Origin	Plasma membrane	Intracytoplasmic vesicles
2. Mechanism of formation	Blebbing of cytoplasmic membrane	Fusion of the endoplasmic membrane of intracellular vesicles with the cytoplasmic membrane
3. Size	0–1–1 um	20–100 nm
4. Variance	Heterogenous	Homogeneous
5. Surface markers	CD40	Tetraspanins
6. Membrane components	Enriched in lipids	Enriched in lipids
7. Contents	Proteins, mRNA, miRNA, lipids	Proteins, mRNA, miRNA, lipids
8. Mechanism of release	1. Ca^{2+}-dependent 2. Stimuli-dependent	1. ESCRT-dependent 2. Tetraspanin-dependent 3. Ceramide-dependent 4. Stimuli-dependent

bFGF basic fibroblast growth factor; *SMCs* smooth muscle cells; *ESCRT* endosomal sorting complexes required for transport.

Exosomes, first described by Pan and Johnstone in 1983, are a member of the family of extracellular vesicles with a diameter of 40–120 nm. When first described, exosomes were thought to carry cellular waste. Recently, it was proven that they carry genetic materials, for example, mRNA and miRNA, which play an important role in the cell–cell cross talk and in cellular function.

Exosome synthesis, trafficking, and biofunction are controlled by a complex network of transport machinery (ESCRT-0, I, II, and III). More than 900 different molecules were isolated from exosomes, as reported by ExoCarta. In MSCs, exosomes carry cytokines and growth factors, such as interleukin-6, interleukin 10, TGFB, HGF, and VEGF. They also carry pro-angiogenic factors, such as miRNA-125a, miRNA-30b, miRNA-30c, miRNA-424, miRNA-150, and ket-7f. Tissue repair proteins such as MMP-9 and extracellular matrix metalloproteinase inducer.

Exosomes are extremely heterogenous both in membrane composition and in the content of bioactive molecules. This heterogeneity is governed both by the cell of origin and the stimulus of release. Preconditioning of MSCs can modify the exosome content according to the microenvironment.

Factors affecting exosome content:

Exosome cargo content varies according to a number of factors:

1. Cell lineage
2. Cell source
3. Microenvironment: changing the milieu of the cells greatly influences their exosomal content. Thus, directed microenvironment and culture conditions can enhance exosomal content of certain angiogenic, anti-angiogenic, neural, or other regenerative bioactive molecules.

4.6.2 Paracrine action

Haynesworth et al. were the first to report that MSCs secrete a number of growth factors, chemokines, and regulatory proteins that can participate in tissue regeneration. MSCs secretome includes a variety of cytokines, growth factors, chemokines metalloproteinases, and bioactive molecules. These molecules can be secreted directly into the microenvironment (Table 4.3).

Together with transdifferentiation, cell fusion is another phenomenon of stem cell plasticity. Cell fusion is a natural process implicated in normal development, immune response, tissue formation, and physiological regeneration, where merging of two cells takes places to form a

Table 4.3 Mesenchymal stem cell secretome.

Type	Molecule	Function
Growth factors	TGFB	1. Treg activation 2. Suppression of DC maturation 3. Suppression of T17 generation
	HGF	
	VEGF	1 Angiogenesis 2 Neurogenesis
Interleukins	1. IL-1Ra	1. M2-like macrophage activation 2. Inhibition of B cell activation
	2. LIF	1. Inhibition of T17 differentiation 2. Induction of Treg proliferation
Chemokines	mpCCL2	Prevents T17 chemotaxis and activation
Prostaglandins	PGE2	1. Induces macrophage production of IL-10 2. Promotes Macrophage differentiation
Metabolic molecules	Gal1	1. Inhibits proliferation of CD4 and CD8 T cells
	Gal9	Induces apoptosis of Th17 and Th1 cells

syncytium, a new hybrid cell expressing a new phenotype. Several reports have noted the contribution of fusion to natural repair with examples such as one-marrow stem cells fusing with several cell types, especially hepatocytes. Cell fusion is a major source of bone marrow–derived hepatocytes under hepatic degenerative conditions. Studies of patients undergoing gender-mismatched bone marrow transplantation have previously documented the presence of donor-derived mature cells in almost all body organs. This was explained by transdifferentiation of bone marrow–derived stem cells into mature body cells. However, one study documented the presence of rare tetraploid neurons and hepatocytes showing (XXXY), a finding that introduces the process of fusion as another possibility of stem cell contribution to regeneration.

Cell fusion can be a mode of induction of regeneration and correction of autosomal recessive disorders.

4.7 Mitochondrial transfer

Mitochondrion, a membrane-bound cell organelle, is considered the power generator of the cell. It is responsible for converting oxygen and nutrients into ATP, which is the energy "currency" inside the cells.

Mitochondria play an important role in oxidative phosphorylation, prevention of apoptosis, and ATP generation. Mitochondrial transfer is known in many somatic cells and fibroblasts.

Since MSCs were reported to reduce the amount of reactive oxygen species (ROS) in tissues, it was suggested that mitochondrial transfer is one of the regenerative mechanisms implicated in MSC-based therapy. Mitochondrial transfer from stem cells to degenerative or stressed cells is a mode of cellular repair, which was documented in MSCs in many studies. Coculture of MSCs with mitochondrial-deprived cells was found to induce restoration of mitochondria to these cells.

4.7.1 Inducers of transfer

Mitochondrial transfer is triggered by the presence of ROS in the microenvironment as well as the presence of mitochondrial damage. Damaged mitochondria release mitochondrial DNA and products called damage-associated molecular patterns, which induce mitochondrial transfer. CD38 upregulation also stimulates mitochondrial transfer.

4.7.2 Mechanisms

Mitochondria are transferred between cells through one of the following mechanisms:

1. Tunneling nanotubes
2. Microvesicle formation
3. Gap junctions
4. Cell fusion
5. Direct uptake of mitochondria from the medium

4.7.3 Effects

MSCs present in areas of stress secrete depolarized mitochondria for their protection through transcellular mitophagy. Damaged cells release mitochondria in the environment, which are engulfed and degraded by MSCs, causing induction of cytoprotective enzyme HO-1 and mitochondrial biogenesis.

4.7.4 Variability in mitochondrial transfer according to mesenchymal stem cell source

Paliwal et al. (2018) demonstrated the differential mitochondrial transfer potential across MSC sources. They found that dental pulp and Wharton jelly–derived MSCs exert more potent mitochondrial transfer potential than bone marrow and adipose tissue–derived MSCs. This was proven through a quantitative comparison between the ROS in tissues.

4.8 Microenvironment modulation

Through their antiinflammatory, antimicrobial, and immunomodulatory functions, MSCs are able to change the microenvironment from a degenerative to a regenerative setting. Changing the microenvironment milieu from pro- to antiinflammatory promotes angiogenesis, cell proliferation, and reduces cell apoptosis. This effect is mediated by altering the function of immune cells either through cell–cell contact, paracrine, or exosome mediated.

A. Antiinflammatory effects:

The antiinflammatory effects involved in regenerative function of stem cells are:

1. Reduction of neutrophil migration and infiltration
2. Reduction of pro-inflammatory cytokine levels: IL-1, INF-gamma, IL-6, macrophage inflammatory protein-1. This is mostly mediated through IL-10 secretion, which inhibits the pro-inflammatory cytokines by reduction of hemoglobin-induced oxidative tissue damage through positive feedback of CD163.
3. Increased granulocyte colony stimulating factor level leading to enhanced granulocyte production and function.
4. Inhibition of pyroptosis, caspase-dependent cell death, which is accompanied by the release of inflammasomes inducing marked inflammatory response. This is mediated through IL-10 secretion.

B. Antimicrobial functions:

MSCs exert their antimicrobial functions through a number of mechanisms. They have direct bactericidal activity, which is present spontaneously in vitro. Moreover, MSC-related soluble factors inhibit the growth of a number of bacteria in addition to drug-resistant bacteria. Lastly, MSCs augment the innate immune response, inducing neutrophil extracellular trap formation and bacterial phagocytosis. The antimicrobial activity of MSCs can also be mediated through mitochondrial transfer.

C. Immunomodulatory functions:

Immunomodulatory function of MSCs targets almost all arms of the immune system: T-lymphocytes, B-lymphocytes, NK lymphocytes, macrophages, and dendritic cells.

1. T cells:

   MSCs inhibit the proliferation and activation of T cells (both CD4 and CD8 T cells) and cause shift of T cell phenotype from pro-inflammatory Th1 to antiinflammatory Th2 with all associated cytokine profile shift. MSCs also induce the formation and expansion of CD4 + CD25-FoxP3- Tregs. Lastly, MSCs can suppress T cell activation and induce T cell apoptosis through PD-L1 secretion.
2. B cells:

   MSCs effects on B-cells are largely mediated through direct cell-to-cell contact, reducing plasmablast formation and inducing the production of Bregs. Bregs exert their immunotolerant and antiinflammatory effects through the production of IL-10. In addition to the direct mechanism, MSCs inhibit B cell proliferation through cytokine routes, namely, the production of IL-1alpha as well as inducing T cell interferon gamma production. IL-1alpha inhibits B cell differentiation, while INF-gamma inhibits B cell proliferation.
3. NK cells:

   MSCs are strong inhibitors of NK cells: both proliferation and cytotoxic activity. This effect is probably mediated through the production of prostaglandin E2, IL-6, TGF-B, and nitric oxide.
4. Macrophages:

   MSCs were found to drive macrophages to polarize into antiinflammatory M2 instead of into pro-inflammatory M1, probably through secretion of IL-1 receptor agonist (IL-1 RA). Another pathway implicated in this shift is through monocyte phagocytosis of MSCs, which causes monocytes to turn to type 2 immunosuppressive phenotype. Phagocytosis of MSC caused CD14 + CD16- mature monocytes to polarize toward a CD14+ + CD16 + CD206 + immune regulatory phenotype with strong antiinflammatory properties secreting IL-10 and PDL-1 and enhanced ability to induce CD4 + CD25 + T-reg formation.
5. Dendritic cells:

   DCs cultured with MSCs in vitro show reduced ability to induce antigen-specific T cell proliferation and MHC class II expression. Type 1 DCs secrete less TNF-alpha and type 2 DCs secrete more IL-10 in the presence of MSCs.

Also, MSCs were found to induce differentiation of CD34 + HSCs to regulatory DCs. MSCs also suppress the migration and maturation of DCs.

4.9 Conclusion

Stem cells induce tissue regeneration through a number of mechanisms. In most cases, all these mechanisms function in harmony. However, in some cases, one of these arms takes the upper hand. Understanding these mechanisms can lead to enhancement of stem cell–based regeneration.

Further reading

Baglio SR, Pegtel DM, Baldini N. Mesenchymal stem cell secreted vesicles provide novel opportunities in (stem) cell-free therapy. Front Physiol 2012;3:359.

Baraniak P, McDevitt T. Stem cell paracrine actions and tissue regeneration. Regen Med 2010;5(1):121–43.

Camp J, Beckers S, Zegers D, Hul W. Wnt signaling and the control of human stem cell fate. Stem Cell Rev 2013;10:2.

Cortes-Medina L, Pasantes-Morales H, Castrejon A, Picones A, Lara Figueroa C, Luis E, et al. Neuronal transdifferentiation potential of human mesenchymal stem cells from neonatal and adult sources by a small molecule cocktail. Stem Cell Int 2019;2019:13 | Article ID 7627148.

Eda A, Osawa H, Satoh K, Yanaka I, Kihira K, Ishino Y, et al. Aberrant expression of CDX2 in Barrett's epithelium and inflammatory esophageal mucosa. J Gastroenterol 2003;38:14–22.

Graf T, Enver T. Forcing cells to change lineages. Nature. 2009;462(7273):587–94.

Gramignoli R, Sallustio F, Widera D, Raschzok N. Tissue repair and regenerative mechanisms by stem/progenitor cells and their secretome. Front Med 2019;. Available from: https://doi.org/10.3389/fmed.2019.00011 published: 31 January 2019.

Hatzistergos K, Quevedo H, Oskouei B, Hu Q, Feigenbaum G, Margitich I, et al. Bone marrow mesenchymal stem cells stimulate cardiac stem. Circ Res 2010;107 (7):913–22.

Haynesworth SE, Baber MA, Caplan AI. Cytokine expression by human marrow-derived mesenchymal progenitor cells in vitro: effects of dexamethasone and IL-1α. J Cell Physiol 1996;166(3):585–92.

Ieda M, Fu JD, Delgado-Olguin P, Vedantham V, Hayashi Y, Bruneau BG, et al. Direct reprogramming of fibroblasts into functional cardiomyocytes by defined factors. Cell. 2010;142(3):375–86.

Iyer S, Rojas M. Anti-infl ammatory effects of mesenchymal stem cells: novel concept for future therapies. Expert Opin Biol Ther 2008;8(5):569–81.

James R, Kazenwadel J. Homeobox gene expression in the intestinal epithelium of adult mice. J Biol Chem 1991;266:3246–51.

Johnstone RM, Adam M, Hammond JR, Orr L, Turbide C. Vesicle formation during reticulocyte maturation. Association of plasma membrane activities with released vesicles (exosomes). J Biol Chem 1987;262(19):9412–20.

Korinek V, Barker N, Moerer P, van Donselaar E, Huls G, Peters PJ, et al. Depletion of epithelial stem-cell compartments in the small intestine of mice lacking Tcf-4. Nat Genet 1998;19:379–83.

Lai RC, Yeo WY, Lim SK. Mesenchymal stem cell exosomes. Seminars in cell & developmental biology, Vol. 40. Academic Press; 2015. p. 82–8.

Li H, Shen S, Fu H, Wang Z, Li X, Sui X, et al. Immunomodulatory functions of mesenchymal stem cells in tissue engineering. Stem Cell Int 2019;2019(9671206):18.

Li W, Yu W, Quinlan J, Burke Z, Tosh D. The molecular basis of transdifferentiation. J Cell Mol Med 2005;9(3):569–82.
Lismaa S, Kaidonis X, Nicks A, Bogush Nm Kikuchi K, Naqvi N, Harvey R, et al. Comparative regenerative mechanisms across different mammalian tissues. NPJ Regen Med 2018;3:6.
Loffredo F, Steinhauser M, Gannon J, Lee R. Bone marrow-derived cell therapy stimulates endogenous cardiomyocyte progenitors and promotes cardiac repair. Cell Stem Cell 2011;8(4):389–98.
Lu H, Wang F, Mei H, Wang S, Cheng. Human adipose mesenchymal stem cells show more efficient angiogenesis promotion on endothelial colony-forming cells than umbilical cord and endometrium. Stem Cell Int 2018;2018(7537589):15.
Luo L, Hu D, Yin J, Xu R. Molecular mechanisms of transdifferentiation of adipose-derived stem cells into neural cells: current status and perspectives. Stem Cell Int 2018;2018(5630802):14.
Marlein C, Piddock R, Mistry J, Zaitseva L, Hellmich C, Horton R, et al. CD38-Driven mitochondrial trafficking promotes bioenergetic plasticity in multiple myeloma. Cancer Res 2019;. Available from: https://doi.org/10.1158/0008-5472.CAN-18-0773.
Martínez-Losa M. Paracrine mechanisms of stem cell reparative and regenerative actions in the heart. Adv Exp Med Biol 2011;713:161–75.
Miranda F, Varas-Godoy M, Khoury M. Harnessing the angiogenic potential of stem cell-derived exosomes for vascular regeneration. Stem Cell Int 2016;2016(3409169):11.
Mutoh H, Hakamata Y, Sato K, Eda A, Yanaka I, Honda S, et al. Conversion of gastric mucosa to intestinal metaplasia in Cdx2-expressing transgenic mice. Biochem Biophys Res Comm 2002;294:470–9.
Mutoh H, Sakurai S, Satoh K, Osawa H, Hakamata Y, Takeuchi T, et al. Cdx1 induced intestinal metaplasia in the transgenic mouse stomach: comparative study with Cdx2 transgenic mice. Gut 2004;53:1416–23.
Owen R, Gough J. A predictive computational framework for direct reprogramming between human cell types. Nat Genet 2016;48(3):331–5.
Paliwal S, Chaudhuri, Agrawal A, Mohanty S. Regenerative abilities of mesenchymal stem cells through mitochondrial transfer. J Biomed Sci 2018;25:31.
Peruzzaro S, Andrews M, Al-Gharaibeh A, Pupiec A, Resk M, Story D, et al. Transplantation of mesenchymal stem cells genetically engineered to overexpress interleukin-10 promotes alternative inflammatory response in rat model of traumatic brain injuy. J Neuroinflammation 2019;16:2.
Qin Y, Guan J, Zhang C. Mesenchymal stem cells: mechanisms and rolein bone regeneration. Postgrad Med J 2014;90:643–7.
Qu C, Brohlin M, Kingham P, Kelk P. Evaluation of growth, stemness, and angiogenic properties of dental pulp stem cells cultured in cGMP xeno-/serum-free medium. Cell Tissue Res 2020;380:93–105.
Redondo-Castro E, Cunningham C, Miller J, Martuscelli L, Aoulad-Ali S, Rothwell N, et al. Interleukin-1 primes human mesenchymal stem cells towards an anti-inflammatory and pro-trophic phenotype in vitro. Stem Cell Res Ther 2017;8:79.
Reinecke H, Minami E, Zhu W, Laflamme M. Cardiogenic differentiation and transdifferntiation of progenitor cells. Circ Res 2008;103(10):1058–71.
Rezaie J, Heidarzadeh M, Hassanpour M, Amini H, Shokrollahi E, Ahmadi M, et al. The angiogenic paracrine potential of mesenchymal stem cells. Update on Mesenchymal and Induced Pluripotent Stem Cells 2019;.
Rodriguez L, Mohammadipoor A, Alvarado L, Kamucheka R, Asher A, Cancio L, et al. Preconditioning in an inflammatory milieu augments the immunotherapeutic function of mesenchymal stromal cells. Cells 2019;2019(8):462.

Ruhnke M, Ungefroren H, Nussler A, Martin F, Brulport M, Schormann W, et al. Differentiation of in vitro-modified human peripheral blood monocytes into hepatocyte-like and pancreatic islet-like cells. Gastroenterology 2005;128:1774–86.

Saldana L, Bensjamar F, Valles G, Mancebo F, Garcia-Rey E, Vilaboa N. Immunoregulatory potential of mesenchymal stem cells following activation by macrophage-derived soluble factors. Stem Cell Res Ther 2019;10:58.

Seifert AW, Monaghan JR, Smith MD, Pasch B, Stier AC, Michonneau F, et al. The influence of fundamental traits on mechanisms controlling appendage regeneration. Biol Rev Camb Philos Soc 2012;87:330–45.

Sharma P, McQuaid K, Dent J, Fennerty MB, Sampliner R, Spechler S, et al. A critical review of the diagnosis and management of Barrett's esophagus: the AGA Chicago workshop. Gastroenterology 2004;127:310–30.

Stoltz JF, de Isla N, Li YP, Bensoussan D, Zhang C, Huselstein C, et al. Stem cells and regenerative medicine: myth or reality of the 21th century. Stem Cell Int 2015;2015:19 |Article ID 734731.

Takeuchi R, Katagiri W, Endo S, Kobayashi T. Exosomes from conditioned media of bone marrow-derived mesenchymal stem cells promote bone regeneration by enhancing angiogenesis. PLOS ONE14 2019;11).

Taléns-Visconti R, Bonora A, Jover R, Mirabet V, Carbonell F, Castell J, et al. Hepatogenic differentiation of human mesenchymal stem cells from adipose tissue in comparison with bone marrow mesenchymal stem cells. World J Gastroenterol 2006;28(12(36)):5834–45.

Tao H, Han Z, Han ZC, Li Z. Proangiogenic features of mesenchymal stem cells and their therapeutic applications. Stem Cell Int 2016;2016(1314709):11.

Tosh D, Slack JMW. How cells change their phenotype. Nat Rev Mol Cell Biol 2002;3:187–94.

Vega ME1, Giroux V, Natsuizaka M, Liu M, Klein-Szanto AJ, Stairs DB, et al. Inhibition of notch signaling enhances transdifferentiation of the esophageal squamous epithelium towards a Barrett's-like metaplasia via KLF4. Cell Cycle 2014;13(24):3857–66. Available from: https://doi.org/10.4161/15384101.2014.972875.

Wallis D, Hamblen M, Zhou Y, Venken KJ, Schumacher A, Grimes HL, et al. The zinc finger transcription factor Gfi1, implicated in lymphomagenesis, is required for inner ear hair cell differentiation and survival. Development 2003;130:221–32.

Wang J, Ren H, Yuan X, Ma H, Shi X, Ding Y. Interleukin-10 secreted by mesenchymal stem cells attenuates acute liver failure through inhibiting pyroptosis. Hepatology Res 2018;48:E194–202.

Watt S, Gullo F, van der Garde M, Markeson D, Camicia R, Khoo, et al. The angiogenic properties of mesenchymal stem/stromal cells and their therapeutic potential. Br Med Bull 2013;108(1):25–53.

Wells J, Watt F. Diverse mechanisms for endogenous regeneration and repair in mammalian organs. Nature 2018;557:7705.

Whitby DJ, Longaker MT, Harrison MR, Adzick NS, Ferguson MW. Rapid epithelialisation of fetal wounds is associated with the early deposition of tenascin. J Cell Sci 1991;99(Pt 3):583–6.

White I, Sanina C, Balkan W, Hare J. Mesenchymal stem cells in cardiology. Methods Mol Biol 2016;1416:55–87.

Ye J, Su X, Stoltz J, de Isla N, Zhang L. Signalling pathways involved in the process of mesenchymal stem cells differentiating into hepatocytes. Cell Prolif 2015;48:157–65.

CHAPTER 5

Sources of stem cells for regenerative purposes

Contents

5.1 Characteristics of the ideal source of stem cells 57
5.2 Sources of mesenchymal stem cells 58
5.3 Skin mesenchymal stem cells 62
5.4 Periodontal ligament stem cells 62
5.4.1 Sources of hemopoietic stem cells 62
5.5 Cord blood 63
5.5.1 Sources of endothelial cells 63
5.6 Tissue-specific stem cells 64
Further reading 66

Despite the extensive research in regenerative medicine using a variable array of stem cell types extracted from various sources, the most promising stem cell lineage for regenerative purposes is the mesenchymal stem cell (MSC). Hemopoietic stem cells (HSCs), endothelial stem cells, and induced pluripotent stem cells are also efficient candidates for stem cell therapy. Thus, the choice of the stem cell source must be proceeded by choosing the stem cell type.

5.1 Characteristics of the ideal source of stem cells

A source for stem cell harvest must fulfill certain criteria in order to be considered a potential reliable source for stem cells. This source should provide adequate supply of stem cells sufficient for the stem cell therapy procedure. The process of stem cell harvest should be safe with no hazards neither to the donor nor to the cell viability or function. Moreover, the stem cells isolated should be renewable so as to allow frequent sampling if needed.

Stem Cell Therapy
DOI: https://doi.org/10.1016/B978-0-12-821569-2.00006-5

5.2 Sources of mesenchymal stem cells

MSCs are present in all body organs and can be isolated from almost all body tissues. However, practically, clinically usable sources for MSCs are bone marrow, adipose tissue, and cord tissue. Less abundant sources are dental pulp, fetal tissue (placenta and amniotic fluid), synovial fluid and tissue, endometrium, and skin (Table 5.1).

According to the "minimal criteria of mesenchymal stem cell identification" indicated by the International Society for Cellular Therapy, MSCs from various sources share the baseline characteristics of MSCs. However, the stem cell source also dictates additional characteristics that distinguishes MSCs isolated from it. Bone marrow mesenchymal stem cells (BM-MSCs) are regarded as the "golden standard" of MSCs.

Bone marrow is the "reference" source of MSCs. Isolation of MSCs from bone marrow is technically easy with convenient expansion and transdifferentiation protocols. MSCs represent 0.01%–0.001% of bone marrow population. Bone marrow–derived MSCs are genetically stable for numerous passages and are well tolerated when injected both systemically and locally. The main disadvantage of bone marrow as a source for MSCs is that its acquisition is somewhat invasive, requiring sedation or anesthesia (Fig. 5–1).

Peripheral blood is another source of MSCs, the source of which is disputed. The fact that peripheral blood mesenchymal stem cells (PB-MSCs) show minor biological differences from BM-MSCs can be explained as either postmobilization adaptive changes or the result of a different origin. PB-MSCs are much more abundant in children, and their incidence declines with age. Circulating MSCs, with or without mobilization, can be collected from PB through leukapheresis procedure. PB-MSCs are biologically identical to BM-MSCs except for the cycling status. Mobilized MSCs are more cycling than those isolated from bone marrow. Thus, PB-MSCs show lower proliferative rate than BM-MSCs. PB-MSCs show the immunophenotypic pattern typical of MSCs and express CD146 and CD140b, which are exclusively expressed by PB-MSCs. PB-MSCs show higher adipogenic differentiation potential than BM-MSCs.

Cord tissue (Wharton jelly) is a rich source of MSCs. UC-MSCs are collected noninvasively as they are obtained from medical waste. The isolation process is easy with large numbers of MSCs. UC-MSCs have shorter doubling time than BM-MSCs and show higher viability at high passage

Table 5.1 Comparison between mesenchymal stem cells from different sources.

	Bone marrow	Peripheral blood	Adipose tissue	Cord blood	Skin	Periodontal ligament
Procedure	Invasive	Minimally invasive	Invasive	Discarded tissue	Invasive	Discarded tissue
Painful procedure	++	−	+	−	+	−
Processing	None	None	Difficult	Difficult	Difficult	Difficult
Expense	++	++++	++	+	+	+
MSC %	++	+	+++++	++	++	++
Expansion potential						
Proliferation potential	+++	++	++++	++++	++	++
Doubling time				Shorter than BM		
Immunophenotyping:						
CD90	+	+	+	+	+	+
CD73	+	+	+	+	+	+
CD105	+	+	+	+	+	+
CD44	+	+	+	+	+	+
CD271	+	+	+	+	+	+
CD34	−	−	+	−	−	−
CD45	−	−	−	−	−	−
CD14	−	−	−	−	−	−
CD31	−	−	−	−	−	−
CD146	−	+	−	−	−	−
CD140b	−	+	−	−	−	−
CD 166	−	−	−	−	−	+

(Continued)

Table 5.1 (Continued)

	Bone marrow	Peripheral blood	Adipose tissue	Cord blood	Skin	Periodontal ligament
Plasticity:						
1. Osteogenic Differentiation	+++	++	++	++	++	++
2. Chondrogenic Differentiation	+++	++	++	++	++	++
3. Adipogenic Differentiation	+	++	++	++	++	++
4. Other lineages:						
Neural	+	±	±	±	±	±
Stability in passages	+	+	+	++	+	+
Paracrine activity						
IFN			++			
TGF			++			
IGF			++			
SCF	++					
HGF	++					
Immunomodulatory activity	+	+	+++	+	+	+

− no, + mild, ++ moderate, +++ marked.

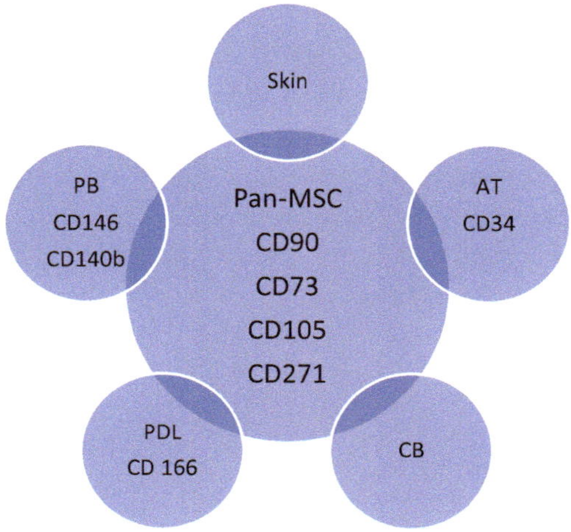

Figure 5–1 Immunophenotyping differences between mesenchymal stem cells from different sources. *AT*, adipose tissue; *PDL*, periodontal ligament; *PB*, peripheral blood; *CB*, cord blood.

numbers. Paracrine function of UC-MSCs is potent, secreting high concentrations of immunomodulatory cytokines: IL-10, IL-8, TGF- β2, and HGF.

MSCs can also be isolated from Wharton jelly and placental tissue, with similar biological characteristics. Wharton jelly MSCs show higher percentage of cells in the G0/G1 phase. Placental MSCs show more powerful immunosuppressive function than MSCs from other sources.

Adipose tissue MSCs are emerging as a good alternative to bone marrow MSCs. MSCs in adipose tissue are localized in the stromal vascular fraction (SVF). They are a heterogenous population which express all the defining characteristics of MSCs including potential for osteogenic, chondrogenic and adipogenic differentiation; and they also show variable plasticity towards other lineages.

Adipose tissue contains a mixed stem cell population: MSC, endothelial cells, and HSCs, in addition to adipose stem cells (ASCs). All these populations exhibit variable regenerative and transdifferentiation potentials. On the whole, adipose tissue–derived stem cells show more potent neural differentiation potential as compared to the higher osteogenic and chondrogenic potential of bone marrow–derived MSCs.

Adipose tissue harbors a higher incidence of stem cells per gram tissue than bone marrow. It is estimated that the stromal vascular fraction contains 3% stem cells, as compared to 0.001% to 0.01% in bone marrow.

Thus, about 0.5×10^4 to 2×10^5 stem cells can be isolated from a gram of adipose tissue, a yield 100–500 fold higher than that of the bone marrow. Adipose-derived stem cells show higher proliferative potential than those derived from bone marrow or Wharton jelly.

ASCs also express peculiar paracrine activity with higher basic fibroblast growth factor (BFGF), interferon (IFN)-gamma, and insulin-like growth factor 1 secretion and more potent immunomodulatory function.

Interestingly, immunophenotyping of ASCs shows positive CD34, with negative CD45 and CD 31 and the expression of MSC markers according to the minimal criteria of ISCT (CD90, CD73, CD105, and CD44).

5.3 Skin mesenchymal stem cells

MSCs isolated from different regions of the skin (SMCs) show similar biological characteristics to the BM-MSCs, with minor differences. SMCs require a longer time to grow in primary culture, but in subsequent passages, their growth rate is higher than BM-MSCs. They also show different paracrine function with higher secretion of proliferative cytokines than BM-MSCs.

5.4 Periodontal ligament stem cells

Periodontal ligament stem cells are isolated from discarded tissue following dental procedure.

Their isolation is relatively easy, with a high proliferative potential and lineage differentiation ability.

5.4.1 Sources of hemopoietic stem cells

Although HSCs represent the benchmark for all subsequently characterized stem cells, they may not be the best candidate for regenerative medicine. In the setting of hemopoeitic stem cell transplantation (HSCT) for hematological reconstitution, peripheral blood hemopoietic cells (PB-HSCs) are easier to isolate in large numbers with less expenses and invasiveness. However, in a regenerative setting, few studies compared the regenerative function of PB and bone marrow hemopoietic cells (BM-HSCs). PB-HSCs show the same plasticity, with faster engraftment than BM-HSCs.

Table 5.2 Conditions associated with increased circulating hemopoietic stem cell.

Noninduced	Induced
Tissue injury: coronary heart diseaseAutoimmune diseases:Celiac Disease	Cytokines: G-CSF, GM-CSFChemotherapy: Endoxan

Bone marrow is the classic source of HSCs. Bone marrow harvest is the classic method of obtaining BM-HSCs. Bone marrow harvest is an invasive technique done under general anesthesia.

Peripheral blood: HSC circulation between bone marrow, peripheral blood, and other extramedullary hemopoietic organs, such as the liver and spleen, is a central part of hematopoiesis during embryonic life. In postnatal life, HSCs continue to traffic; the explanation of this circulation is not clear. However, it may be a part of the regenerative rather than the hemopoietic function of HSCs. HSCs injected intravenously home to the bone marrow, whereas BM-HSCs are mobilized to the blood in response to a number of physiological stimuli. This is exaggerated by the administration of cytokines or chemotherapy, a phenomenon referred to as "mobilization" (Table 5.2).

5.5 Cord blood

Cord blood was introduced as the third source of HSCs after the nuclear accident of Chernobyl in the 1980s. Research proved the ability of umbilical cord blood hemopoeitic stem cells (UCB-HSCs) to reconstitute hemopoiesis. This was followed by the introduction of UCB unit banking, with more than 600,000 stored units in banks worldwide.

With the advent of the regenerative medicine era, UCB stem cells were introduced as regenerative units (Table 5.3).

5.5.1 Sources of endothelial cells

Endothelial progenitors, also known as lectin-positive progenitors, are sometimes used for regenerative purposes, mainly in revascularization. Endothelial stem cells and progenitors can be isolated from bone marrow, peripheral blood, cord blood, Wharton jelly, and adipose tissue. Endothelial

Table 5.3 Comparison between sources of hemopoietic stem cell.

	Bone marrow	Peripheral blood	Umblical cord
Availability	Renewable	Renewable	Nonrenewable
Technique	Invasive	Less invasive	Noninvasive
Collection/risk	Surgery-related severe adverse events 1/100	Apheresis related severe adverse events1/1000	Post delivery None
Number of HSC (Relative number of CD34 + cells)	50 ×	10 ×	1 ×
Cycling state	Noncycling (G0−1)	Cycling (G1−2)	Noncycling (G0−1)
Cost	++	+	+

progenitors show variable expressions of CD 146, CD 34, CD 31, CD 90, and VEGR2.

Endothelial stem cells can be isolated from bone marrow, adipose tissue, umbilical cord Wharton jelly. They are also found in peripheral blood and various human tissues, such as cardiac tissue. However, isolation of ECs for regenerative purposes is usually done from bone marrow, adipose tissue, or Wharton jelly.

Endothelial stem cells probably share common ancestor with other bone marrow stem cells such as HSCs or MSCs. Thus, they can be obtained from transdifferentiation of these stem cells.

5.6 Tissue-specific stem cells

All body tissues contain tissue-resident stem cells responsible for the physiological turnover and wear and tear of the tissues. Tissue-specific stem cells can be obtained by tissue biopsy; thus, this is rarely a feasible source for regenerative purposes. An important exception for this concept is the highly complex cells, which are difficult to be obtained by transdifferentiation in-vitro such as hair follicle stem cells or melanocytes.

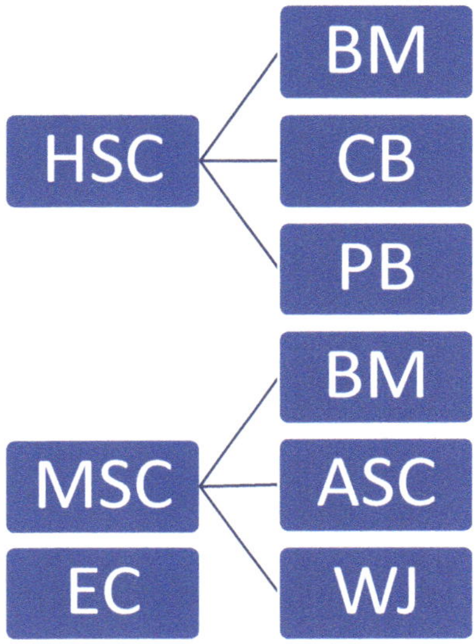
HSC
BM
CB
PB
MSC
BM
ASC
WJ
EC

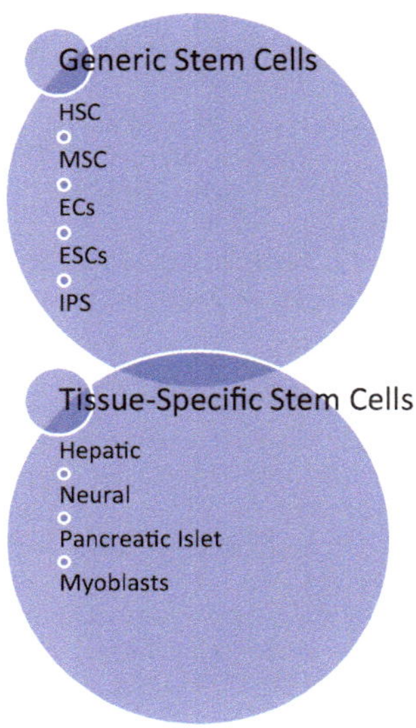
Generic Stem Cells
HSC
MSC
ECs
ESCs
IPS
Tissue-Specific Stem Cells
Hepatic
Neural
Pancreatic Islet
Myoblasts

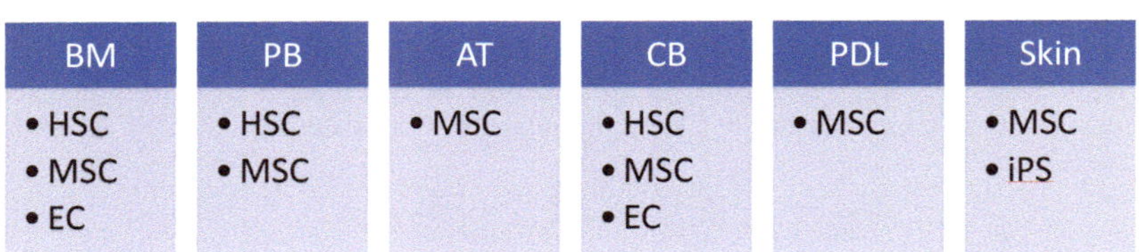
BM
• HSC
• MSC
• EC
PB
• HSC
• MSC
AT
• MSC
CB
• HSC
• MSC
• EC
PDL
• MSC
Skin
• MSC
• iPS

HSC hemopoeitic stem cell
BM bone marrow
CB cord blood
MSC mesenchymal stem cell
ASC adipose stem cell
WJ Wharton jelly
ECs endothelial cells
ESCs endothelial stem cells
IPS induced pluripotent stem cell

Further reading

Baer P, Geiger H. Adipose-derived mesenchymal stromal/stem cells: tissue localization, characterization, and heterogeneity. Stem Cell Int 2012;2012 Article ID 812693, 11 pages.

Berebichez-Fridman R, Montero-Olvera P. Sources and clinical applications of mesenchymal stem cells. Sultan Qaboos Univ Med J 2018;18(3):e264–77.

Berebochez-Fridman R, Montero-Olvera P. Sources and clinical applications of mesenchymal stem cells. State-of-the-art review Stem Cells Int 2018;2018|Article ID 8429042|. Available from: https://doi.org/10.1155/2018/8429042.

Charbord P. Bone marrow mesenchymal stem cells: historical overview and concepts. Hum Gene Ther 2010;21(9):1045–56. Available from: https://doi.org/10.1089/hum.2010.115.

Chen G, Yue A, Ruan Z, Yin R, Wang R, Ren Y, et al. Comparison of biological characteristics of mesenchymal stem cells derived from maternal-origin placenta and Wharton's jelly. Stem Cell Res Ther 2015;6:228.

Chong MS, Ng WK, Chan JK. Concise review: endothelial progenitor cells in regenerative medicine: applications and challenges. Stem Cell Transl Med 2016;5(4):530–8. Available from: https://doi.org/10.5966/sctm.2015-0227 Epub 2016 Mar 8. PMID: 26956207; PMCID: PMC4798740.

Chu D, Phuong T, Tien N, Tran D, Minh L, Thanh V, et al. Adipose tissue stem cells for therapy: an update on the progress of isolation, culture, storage and clinical application. J Clin Med 2019;8(7):917.

Garmy-Susini B, Varner JA. Circulating endothelial progenitor cells. Br J Cancer 2005;93:855–8.

Gentile P, Piccinno M, Calabrese C. Characteristics and potentiality of human adipose-derived stem cells (hASCs) obtained from enzymatic digestion of fat graft. Cells 2019; 8(3):282.

Jang J, Lee H, Eom Y, Kang S. Characteristics of adipose tissue-derived mesenchymal stem cells. Blood 2006;108(11) 4251.

Kim JH, Jo CH, Kim HR, Hwang YI. Comparison of immunological characteristics of mesenchymal stem cells from the periodontal ligament, umbilical cord, and adipose tissue. Stem Cell Int 2018;2018:8429042. Available from: https://doi.org/10.1155/2018/8429042 PMID: 29760736; PMCID: PMC5901833.

Li C, Wu X, Tong J, et al. Comparative analysis of human mesenchymal stem cells from bone marrow and adipose tissue under xeno-free conditions for cell therapy. Stem Cell Res Ther 2015;6:55.

Liu R, Chang W, Wei H, Zhang K. Comparison of the biological characteristics of mesenchymal stem cells derived from bone marrow and skin. Stem Cell Int 2016; 2016:3658798.

Lotfy A, El-Sherbiny Y, Cuthbert R, Jones E, Badawy A. Comparitive study of biological characteristics of mesenchymal stem cells isolated from mouse bone marrow and peripheral blood. Biomed Rep 2019;165–70.

Mohamed-Ahmed S, Fristad I, Lie SA, et al. Adipose-derived and bone marrow mesenchymal stem cells: a donor-matched comparison. Stem Cell Res Ther 2018;9:168. Available from: https://doi.org/10.1186/s13287-018-0914-1.

Monsuur HN, Weijers EM, Niessen FB, Gefen A, Koolwijk P, et al. Extensive Characterization and Comparison of Endothelial Cells Derived from Dermis and Adipose Tissue: Potential Use in Tissue Engineering. PLoS One 2016;11(11):e0167056.

Panch S, Szymanski J, Savani B, Stroncek D. Sources of hematopoietic stem and progenitor cells and methods to optimize yields for clinical cell therapy. Biol Blood Marrow Transpl 2017;23:1241–9.

Ryu HH, Kang BJ, Park SS, Kim Y, Sung GJ, Woo HM, et al. Comparison of mesenchymal stem cells derived from fat, bone marrow, Wharton's jelly, and umbilical cord blood for treating spinal cord injuries in dogs. J Vet Med Sci 2012;74(12):1617–30. Available from: https://doi.org/10.1292/jvms.12-0065 Epub 2012 Aug 9. PMID: 22878503.

Tal R, Dong D, Shaikh S, Mamillapalli R, Taylor HS. Bone-marrow-derived endothelial progenitor cells contribute to vasculogenesis of pregnant mouse uterus†. Biol Reprod 2019;100(5):1228–37. Available from: https://doi.org/10.1093/biolre/ioy265 PMID: 30601943; PMCID: PMC6497522.

Tomokiyo A, Yoshida S, Hamano S, Hasegawa D, Sugi H, Maeda H. Detection, characterization, and clinical application of mesenchymal stem cells in periodontal ligament tissue. Stem Cell Int 2018;2018:5450768.

Trivanović D, Kocić J, Mojsilović S, Krstić A, Ilić V, Djordjević IO, et al. Mesenchymal stem cells isolated from peripheral blood and umbilical cord Wharton's jelly. Srp Arh Celok Lek 2013;141(3–4):178–86. Available from: https://doi.org/10.2298/sarh1304178t PMID: 23745340.

Wagner W, Wein F, Seckinger A, Frankhauser M, Ulfkrause U, Blake JKJ, et al. Comparative characteristics of mesenchymal stem cells from human bone marrow, adipose tissue, and umbilical cord blood. Exp Hematol 2005;33(11):1402–6.

Yoder MC. Human endothelial progenitor cells. Cold Spring Harb Perspect Med 2012;2(7):a006692. Available from: https://doi.org/10.1101/cshperspect.a006692.

Zimmerlin L, Donnenberg VS, Pfeifer ME, et al. Stromal vascular progenitors in adult human adipose tissue [published correction appears in Cytometry A. 2010 Apr;77(4):406]Cytometry A 2010;77(1):22–30. Available from: https://doi.org/10.1002/cyto.a.20813.

CHAPTER 6

Basic concepts in stem cell therapy: stem cell therapy strategy development

Contents

6.1 Introduction 69
6.2 What is the basic defect? 70
6.2.1 Cellular defects 70
6.2.2 Structural defects 71
6.3 Which cell lineage is the best for repair? Choosing optimum cell lineage 71
6.4 What is the best source for a cell type? Choosing the optimum cell source 74
6.4.1 Hemopoietic stem cells 74
6.4.2 Mesenchymal stem cells 74
6.5 What is the optimum laboratory protocol? 74
6.6 Choosing the optimum route of injection 76
6.7 Factors affecting homing of stem cells 77
6.7.1 Methods to improve stem cell homing 78
6.7.2 Mesenchymal stem cell homing 78
6.8 Optimal timing for cell therapy 79
6.9 Optimum number of cells for injection 79
6.10 Choosing the method of evaluation 80
Further reading 80

6.1 Introduction

Application of stem cell therapeutic strategies in regeneration is a relatively new branch of medicine. Formulation of the basic concepts and protocols is currently a hot topic in regenerative medicine. Similar to other branches of medicine, a few questions have to be answered, and according to the answers, precise stem cell therapy protocols can be formulated.

In this chapter, we will discuss these questions and describe how the answers can be reflected on the regenerative strategy of each case (Fig. 6–1).

Stem Cell Therapy
DOI: https://doi.org/10.1016/B978-0-12-821569-2.00001-6

6.2 What is the basic defect?

Organ damage or functional defects requiring stem cell therapy can be either due to loss of a particular cell type or derangement of macro- or microanatomical structure.

6.2.1 Cellular defects

Numerous functional defects are caused by the loss of a particular cell type. A typical example of the acquired type is the loss of pancreatic beta islet cells leading to diabetes mellitus, where the anatomical and physiological structure of the pancreas is normal, but beta islet cells mass and function are diminished. Another example is Parkinson's disease, where the loss of dopaminergic neurons leads to the characteristic features of the disease.

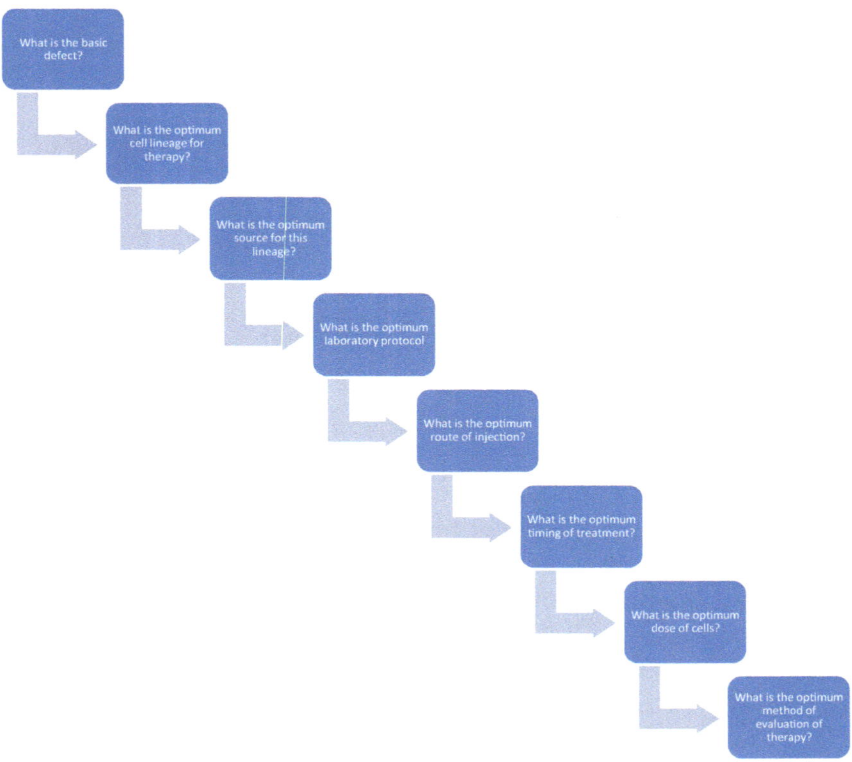

Figure 6–1 Algorithm for regenerative therapy approach.

Congenital defect in protein synthesis can be classified among these defects. These include congenital collagen VII defect in epidermolysis bullosa, enzyme defects in liver metabolic disorders, and amino acidurias.

6.2.2 Structural defects

The second type of damage involves the structural makeup of the organ leading to malfunction. The best example is renal failure, where normal renal anatomy is crucial for function, and cellular replacement cannot induce improvement in function.

A few cases are a mixture of both types, for example, liver cirrhosis, where the pathology involves both cells and the matrix structure (Table 6.1).

6.3 Which cell lineage is the best for repair? Choosing optimum cell lineage

Choosing the cells for repair is dependent on identifying the missing cells. The optimum choice is the same cell lineage, for example, using hepatic stem cells for continuous replacement of hepatocytes in liver cell failure or pancreatic islet cells for diabetes. Using same lineage cells will necessitate the use of allogenic cellular therapy (Fig. 6–2).

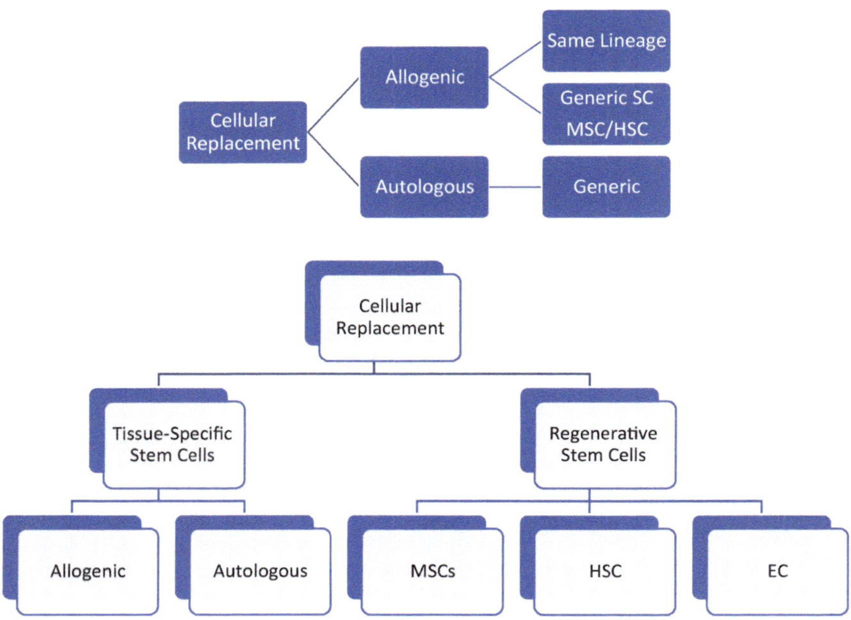

Figure 6–2 Sources for cellular replacement in stem cell therapy.

Table 6.1 Types of defects in degenerative diseases.

Cellular defect		Structural defect		Combined	
Disease	**Type of cells lost**	**Disease**	**Type of structural loss**	**Disease**	**Type of structural loss**
1. Diabetes	Beta islet cells	Renal failure	Nephron	Liver cirrhosis	Hepatic lobule
2. Parkinson's disease	Dopaminergic neurons				
3. Spinal cord injury	Spinal neurons				
4. Retinitis pigmentosa	Retinal pigment cells				
5. Osteoarthritis	Cartilage cells				
6. Avascular Necrosis	Osteocytes				
7. Chronic wounds	Skin epithelium				
8. Azoospermia	Spermatogenic cells				
9. Cardiomyopathy	cardiomyocytes				
10. Idiopathic lung fibrosis	Alveolar epithelium				
11. Optic nerve atrophy	Neurons				
12. Myopathy	Muscle fibers				
13. Ovarian failure	Oocytes				
14. Vitiligo	Melanocytes				
15. Alopecia	Hair follicular stem cells				

The use of tissue-specific stem cells faces two major challenges:

1. *Difficult isolation*: the method of obtaining the stem cell population must be noninvasive and should secure the harvest of a suitable number of the cells sufficient to induce and maintain the regeneration. Tissue-specific stem cells are present in low numbers and in remote anatomical locations in their respective organs. Moreover, these cells are present in vital parts of the body where donation is possible only in cadaveric situation as in the case of cerebral or cerebellar neurons.
2. *Difficult ex vivo expansion* of the cell type: some cell types are almost nonexpandable in vivo, for example, hepatocytes. As these cells must be obtained in large numbers for regenerative purposes, their use in transplantation is limited.
3. Ex vivo expansion of the stem cell population is accompanied by reduction in telomere length. This means that expanded stem cells have a reduced lifespan as compared with nonexpanded stem cells.

The second option for regenerative cell therapy is using the convenient stem cell populations, which are hemopoietic stem cells (HSCs) and mesenchymal stem cells (MSCs) (Table 6.2).

Table 6.2 Comparison between hemopoietic stem cells and mesenchymal stem cells in terms of regenerative potential.

	Hemopoietic stem cell	Mesenchymal stem cell
Sources	• Bone marrow • Cord blood • Leukapheresis	• Bone marrow • Wharton jelly • Adipose tissue • Other sources: amniotic fluid,
Advantages	• Renewable • "Closed-system" isolation available Well-defined immunological and functional characteristics	• Cheap isolation • Easy expansion • Renewable • Nonimmunogenic (no human leukocyte antigen [HLA]-matching needed)
Disadvantages	• Expensive isolation • Difficult Expansion • HLA-matching needed	• No available "closed system" isolation

6.4 What is the best source for a cell type? Choosing the optimum cell source

The commonly used stem cells for therapeutic applications, HSCs and MSCs, can be isolated from a number of sources.

6.4.1 Hemopoietic stem cells

HSCs are classically isolated from bone marrow. They can also be mobilized from bone marrow and collected using leukapheresis. Cord blood is another source of HSCs (Table 6.3).

6.4.2 Mesenchymal stem cells

MSCs hold a great potential as a regenerative tool. Even though they more favorably repair the tissue they originated from, they can be used for repair or all tissue lineages. The heterogeneity of origin of MSCs is reflected in the heterogeneity of their transdifferentiation and paracrine activity crucial for regenerative function (Table 6.4).

6.5 What is the optimum laboratory protocol?

a. Purification:

The best stem cell therapy option is to inject pure cells. However, purification is an additional step that carries the potential of contamination as well as increasing the cost of stem cell therapy. Moreover, the experience in bone marrow transplantation (BMT), which is the

Table 6.3 Sources of hemopoietic stem cells.

	Bone marrow	Cord blood	Leukapheresis
Technique	Invasive	Noninvasive	Minimally invasive
Cost	Inexpensive	Expensive	Expensive
Renewable	Renewable	Nonrenewable	Renewable
Number of cells	High	Limited	High
• Quality of cells:	+ + +	+	+ + + +
• Function	+ + +	+ + + +	+ +
• Self-renewal	+ + +	+	+ + + +
• Immune competence			

Table 6.4 Functional differences between mesenchymal stem cell sources.

	Bone marrow	Adipose tissue	Cord blood and tissues	Other sources (e.g., dental pulp)
Frequency	+ + +	+ + + +	+ +	+ +
Proliferation	+ + +			
Doubling time	40–60 h	45 h	24 h	
Differentiation				
Plasticity:	+ + +	+ + +	+ + +	+ + +
• Adipogenic				
• Chondrogenic				
• Osteogenic				
• Neural				
• Other				
Angiogenic function				
Paracrine function				
Immunomodulatory function				
Senescence	+	+	-	+

practical example of tissue regeneration, does not follow the purification protocols except in specific situations.

Therefore, purification should be restricted to the following conditions:

1. The target organ is very tight-spaced, allowing only a small number of cells; thus, only functional cells have to be injected, for example, ocular space.
2. The target organ is immune-privileged, that is, it normally does not contain immune-competent cells such as the testis. In this case, injecting mononuclear cells, even if autologous, can initiate an immune reaction.
3. The need to eliminate certain cell types, such as lymphocytes in autoimmune diseases.

b. Expansion:

Ex vivo expansion to augment the number of stem cells in the case of a limited source. Expansion here must be before differentiation.

Expansion can reduce the lifespan of stem cells as it is associated with reduced telomere length. Thus, the expansion of the stem cell population can result in faster engraftment and function but may compromise the duration of functioning stem cells, resulting in late graft failure.

c. Differentiation:

Ex vivo differentiation into the target cell lineage can be needed in the following conditions:

1. Transdifferentiation is a two-step process, that is, the cell needed is of different germ layer. In this case, the stem cell should undergo the first step of lineage modification before transplantation.
2. Disrupted microenvironment as in the case of massive hepatic fibrosis, where the normal tissue cells that should secrete inductive cytokines are replaced by fibrous tissues.

Otherwise, stem cells can be injected and left to inductive microenvironmental signals (Tables 6.5 and 6.6).

6.6 Choosing the optimum route of injection

Overall, stem cell therapy for "rejuvenation" is systemic, while degenerative and traumatic indications require local administration.

Injection can be either systemic or local. Systemic injection depends on the homing property of stem cells both in its biologic concept (stem cells will home to their tissue of origin) or in its regenerative concept

Table 6.5 Advantages and disadvantages of pure stem cell population.

Advantages	Disadvantages
1. Small cell volume 2. Reduction of immune activation 3. Adjuvant cells may help cell engraftment through paracrine function (monocytes)	1. Excessive manipulation increasing contamination risk 2. Excessive manipulation compromising cell viability 3. Removal of adjuvant cells, which may help cell engraftment through paracrine function (monocytes)

Table 6.6 Advantages and disadvantages of ex vivo differentiation of stem cells.

	Undifferentiated	Differentiated
In vitro manipulation	Minimal	Excessive
Viability	Optimum	Reduced
Engraftment	Slow	Quick
Paracrine function	More	Reduced
Activation of tissue stem cells	More	Less
Lifespan in vivo	More	Reduced

Table 6.7 Routes of injection in stem cell therapy.

	Systemic administration	Local administration	
		Intra-arterial	Parenchymal
Procedure	Easy	Invasive	Invasive
Engraftment	Homing and engraftment are delayed	Immediate action	Immediate action
Location of injected cells	Less than 1% engraft in the target organ; most of the injected cells are entrapped in the lungs and spleen	100% of cells are delivered to the target organ	100% of cells are delivered to the target organ
Cell retention	Less than 1%	10%–30%	10%–40%

(stem cells home to areas of degeneration and injury). Systemic injection of stem cells can use targeting of the cells to concentrate the cells in the target organ (Table 6.7).

6.7 Factors affecting homing of stem cells

1. Route of administration: The route of stem cell administration determines homing.
2. Timing in tissue injury: Acute tissue injury is associated with cytokine secretion, which induces stem cell homing to the injured sites. This milieu gradually diminishes until the chronic phase begins, where homing is decreased.
3. Administration of exogenous cytokines:

   Exogenous administration of cytokines concomitant with stem cell injection can keep stem cells viable after transplantation until they are completely engrafted. They can also induce vascularization of the prospective homing site to facilitate and promote stem cell engraftment. Cytokines can be administered directly or through genetic modification of the stem cells.

   A. Direct injection of cytokines: to promote engraftment of stem cells. However, cytokines have a short half-life.

   B. Genetic modification of stem cells to overexpress growth factors, such as vascular endothelial growth factor to induce endothelial cell migration and proliferation to promote homing and engraftment of injected stem cells. Transfection of stem cells with IL-10 can induce immune suppression and neural repair. However, the safety of such approaches is still under investigation.

C. Preconditioning of stem cells:

The aim of preconditioning strategies of stem cells is to improve stem cell survival and enhance their regenerative functions: differentiation, proliferation, angiogenesis, antiinflammatory, immune modulatory, and paracrine functions. Generally speaking, tissue environments in need for regeneration are hostile environments with fibrosis, chronic inflammation, and ischemia. Preconditioning is essential to improve survival of stem cells in these damaged tissues. The commonly employed strategies use conditioning medium from functioning mature cells, induction of hypoxia, and using growth factors.

1. Conditioning medium: supernatant of liquid culture of mature cells, which contains extracellular vesicles containing tissue-specific miRNAs and bioactive molecules.
2. Hypoxia: preincubation of stem cells in hypoxia (2%–5% oxygen) activates cellular defense, enhances stem cell viability, and proliferation.
3. Incubation with vascular endothelial growth factor induces cell proliferation and improves survival and reduced cell senescence.

6.7.1 Methods to improve stem cell homing

As the efficiency of stem cell therapy is dependent on homing of sufficient stem cell number to the damaged sites, improvement of stem cell homing is crucial in augmenting the response to stem cell therapy. A number of methods have been explored to enhance HSC homing in the setting of BMT: enhancement of CXCR4 expression, erythropoietin receptor modulation, fucosylation of selectin ligand, and modulation of reactive oxygen species. In addition, coculture of HSCs with MSCs under hypoxic conditions has been found to induce the overexpression of CXCR4.

6.7.2 Mesenchymal stem cell homing

Though MSC homing largely follows the same mechanism as HSC homing, its efficiency is extremely poor. This can be explained by the fact that MSCs are present in all body tissues, so they are diluted over the organs. Another problem is that MSCs administered intravenously can get trapped in the lungs. Thus, concentrating injected MSCs to home to the damaged tissue is a great challenge. A number of approaches are under research: direct administration into target tissue, genetic modification to induce overexpression of CXCR4, cell surface modification of CD44, and in vitro priming using hypoxia to induce TNF and IL-1B expression.

Another approach is to modulate target sites to enhance their ability to recruit injected MSCs. This can be done by inducing an inflammatory environment, using electric stimulation using physiological voltage.

6.8 Optimal timing for cell therapy

The optimal timing for stem cell injection is still unclear. As tissue injury passes through three phases: injury phase (hours), repair phase (days), and remodeling phase (weeks), the microenvironment milieu of cytokines and bioactive molecules changes dramatically along these three phases. During the first two phases, the inflammatory milieu takes the upper hand and will probably compromise the viability of injected stem cells. Going further, some researchers believe that injecting stem cells in this phase will be like adding fuel to the fire. Thus, late repair phase or remodeling phase may be optimum. The duration of each phase as well as its laboratory markers vary according to the tissue.

6.9 Optimum number of cells for injection

Stem cell dose is the most important factor in the efficacy of stem cell therapy. Optimum cell concentration varies according to the organ and its degree of damage and route of administration. The only documented evidence-based information about stem cell dose is the dose used in BMT. Using this information in stem cell regenerative therapy has several drawbacks:

1. HSC is used in BMT, while MSC is used for regenerative medicine;
2. Injection in BMT is systemic, while in MSC, it can be local or systemic;
3. MSCs are a heterogenous population unlike HSCs;
4. Injection in BMT is done once, while in regenerative medicine, it is usually repeated;
5. BMT is accompanied by immune suppression facilitating cell engraftment;
6. In the setting of BMT, bone marrow is the most damaged organ of the body, and the pretransplant conditioning induced a superadded

Table 6.8 Doses in regenerative medicine.

Disease	Organ	Dose
Graft versus host disease (GVHD)	Bone marrow	≥ 2 and $\leq 8 \times 10^6$ /kg
Spinal cord injury	Spinal cord	$0.5-5 \times 10^6$ /kg
Cardiac ischemia	Heart	$1 \times 10^2-2.4 \times 10^9$/kg mononuclear

acute damage to bone marrow, thus inducing chemoattractants in bone marrow.

On the other hand, administration of a large cell dose intravenously will cause blockage of lung capillaries. Direct administration of a large number of cells in the blood supply of the organ cause cell breakdown in the capillaries (Table 6.8).

Stem cell dose required for efficient regeneration is still unclear, taking into considerations that great variations exist according to the targeted organs. Moreover, accumulating evidence suggests that in the regenerative setting, stem cell injection should be done on multiple doses. The number of doses, cell number in each dose, and schedule of the injections are all points yet under investigation. Repeated injections are either using the same cell type and thus induce cumulative regenerative effect or use a cell type as a "receptive" therapy, improving the microenvironment and allowing better engraftment of future cells.

6.10 Choosing the method of evaluation

Determining the method of evaluation of the regenerative effect of administered stem cells is crucial for an objective analysis of the regenerative effect of stem cell therapy and hence improvement of final results.

The method of evaluation should account for two main parameters:

A. Improvement of the initial symptoms (clinical effect)

B. Evaluation of the administered cell homing, engraftment, and function.

Further reading

Florea V, Rieger A, DiFede D, El-Khorazaty J, Natsumeda M, Banerjee M, et al. Dose comparison study of allogeneic mesenchymal stem cells in patients with ischemic cardiomyopathy (the TRIDENT study). Circ Res 2017;121:1279–90.

Golpanian S, Schulman I, Ebert R, Heldman A, Difede D, Yang P, et al. Review and perspective of cell dosage and routes of administration from preclinical and clinical studies of stem cell therapy for heart disease. Stem Cell Tr Med 2016;5:186–91.

Heimfeld S. Bone marrow transplantation: how important is CD34 cell dose in HLA-identical stem cell transplantation? Leukemia, 17. 2003. p. 856–8.
Hénon P. Key success factors for regenerative medicine in acquired heart diseases. Stem Cell Rev Rep 2020;16:441–58.
Liu S, Zhou J, Zhang X, Liu Y, Chen J, Hu B, et al. Strategies to optimize adult stem cell therapy for tissue regeneration. Int J Mol Sci 2016;2117(6):982.
Remberger M, Törlén J, Ringdén O, Engström M, Watz E, Uhlin M, et al. Effect of total nucleated and CD34(+) cell dose on outcome after allogeneic hematopoietic stem cell transplantation. Biol Blood Marrow Transpl 2015;21(5):889–93.
Winkler T, von Roth P, Matziolis G, Mehta M, Perka C, Duda GN. Dose-response relationship of mesenchymal stem cell transplantation and functional regeneration after severe skeletal muscle injury in rats. Tissue Eng Part A 2009;15(3):487–92.
Wysoczynski M, Khan A, Bolli R. New paradigms in cell therapy: repeated dosing, intravenous delivery, immunomodulatory actions, and new cell types. Circ Res 2018;123 (2):138–58.
Yamamoto C, Ogawa H, Fukuda T, Igarashi A, Okumura H, Uchida N, et al. Impact of a low CD34 + cell dose on allogeneic peripheral blood stem cell transplantation. Biol Blood Marrow Transpl 2018;24:708–16.

PART III

Guidelines for stem cell therapy

CHAPTER 7

Stem Cell Therapy: Laboratory guidelines

Contents

7.1 Introduction 85
 7.1.1 Regulatory bodies 86
7.2 Laboratory guidelines 87
 7.2.1 Structural facilities 88
 7.2.2 Personnel qualifications and job description 94
7.3 Quality control team 96
Further reading 96

7.1 Introduction

Stem cell therapy is a relatively new branch of medicine, which is now in the process of foundation of its guidelines of implementation. This process includes establishing detailed steps and standard operating procedures (SOPs) for all phases of stem cell therapy. In the previous chapter, we reviewed the steps for stem cell therapy strategy development. In the present chapter, we will consider in detail the practical steps of stem cell therapy. Despite the great promises stem cell therapy offer, major research and legislation are needed before safe and wide-scale implementation. Stem cell therapy falls in a "gray" zone between research, clinical translation, and innovation.

Although regulatory measures slow the pace of clinical applications and translations, patient safety comes first. Thus, the meticulous generation and application of regulatory measures will ensure patient safety while stem cell therapy is developing and reaching maturity.

Likewise assisted fertilization (in vitro fertilization), stem cell therapy is gaining grounds within the patient populations much faster than within regulatory bodies. Therefore, adopting "balanced" regulatory measures that ensure patient safety while allowing evidence-based research translation is of utmost importance.

Stem Cell Therapy
DOI: https://doi.org/10.1016/B978-0-12-821569-2.00012-0

Regulatory measures include all the requirements and prerequisites of the cellular therapy facility as well as the SOPs of the therapeutic regimen (Fig. 7–1).

7.1.1 Regulatory bodies

Regulatory bodies are arranged in a three-tier model: (1) scientific communities, such as nonprofit organizations of stem cell scientists (the International Society for Stem Cell Research [ISSCR], and International Society for Cell Therapy, [ISCT]), (2) international advisory bodies (the World Health Organization [WHO]), and (3) local governmental regulatory bodies.

Based on their acceptance of stem cell interventions as part of standard medicinal practices, all three pillars of the regulatory system post their input.

ISSCR has issued "Guidelines for stem cell research and clinical translation," delineating the evidence-based facts, state of the art in stem cell research, forecasted advances as well as the probable complications. ISCT has published a number of guidelines for characterization of mesenchymal stem cells (MSC, "Minimal criteria for mesenchymal stem cell identification").

The WHO, an international advisory body, has suggested an action plan according to its constitution. In accordance with this plan, the local governments have issued detailed regulatory laws and acts.

Local regulatory legislations are present in the US, EU, as well as other states. In this chapter, we will describe the basic considerations in regulations as expressed by most regulatory bodies. Otherwise, minimal differences in application of the guidelines should be considered individually.

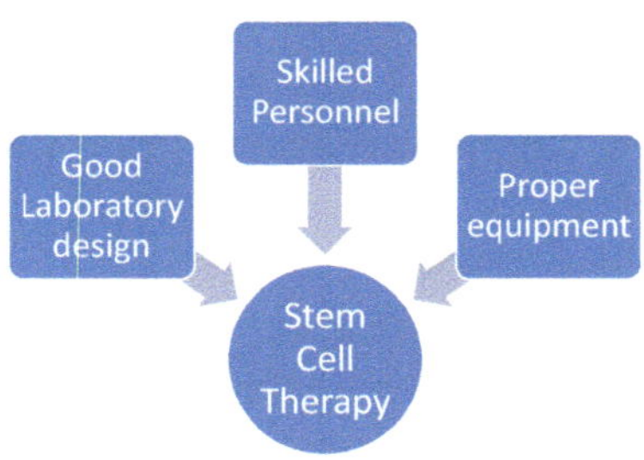

Figure 7–1 Important pillars in stem cell laboratory [four intermingling circles (good laboratory design, proper equipment, skilled personnel, SOPs), with stem cell therapy in the middle].

7.1.1.1 Stem cell therapy types in regard to manipulation

Broadly speaking, stem cell therapeutics are divided into minimally manipulated and effectively manipulated stem cells. Minimal manipulation of stem cells refers to the form of processing that does not alter the biological characteristics of the cells. Examples of minimal manipulation of the cells include centrifugation, washing, freezing, sorting, and the addition of antibiotics or cryopreservatives. On the other hand, effectively manipulated cells are the ones that have undergone cell culture, expansion, in vitro differentiation, or genetic manipulation. The importance of this designation is that each manipulation requires specific qualifications in the facility, equipment, and personnel (Table 7.1).

7.2 Laboratory guidelines

Stem cell therapy is a multidisciplinary approach in which the laboratory is the core facility. Theoretically, stem cell therapy includes embryonic, fetal, as well as adult stem cells. However, practically, only adult stem cell

Table 7.1 Comparison between good tissue practice and good manufacturing practice facilities.

	GTP facility	GMP facility
Purpose	To prevent the introduction, transmission, or spread of communicable diseases by cellular- and tissue-based products	To ensure consistent quality and safety of the products
Type of cellular Therapy	Minimally manipulated cells	Effectively manipulated
Facility	Good tissue practice (GTP)	Good manufacturing practice (GMP)
Environment	No clean room classification	Clean room classification
Quality control	Quality control is required to ensure the sterile handling of products	Quality control is required to 1. Ensure sterility of the product 2. Validate the characterization and safety of the final product

therapy is referred to in most regulatory guidelines in clinical scale therapies. Exceptions are the clinical application of induced pluripotent stem cells in Japan.

The consortium of the laboratory facility includes structural facilities, technical facilities, personnel qualifications, stem cell handling SOPs, as well as quality control systems.

7.2.1 Structural facilities

FDA guidelines for architectural specifications of cell therapy labs are consistent with either good tissue practice (GTP) for minimally manipulated cells or good manufacturing practice (GMP) for effectively manipulated cells.

7.2.1.1 Good tissue practice

GTP requirements are designed to prevent the introduction, transmission, or spread of communicable diseases by cellular- and tissue-based products. They also ensure that these products do not become contaminated during processing. Thus, these requirements are applied for all steps in the process of cellular preparation and begin by donor screening, testing, cell recovery, processing, storage, packaging, and delivery.

GTP facilities should be of suitable size and construction and should be divided into separate areas of adequate size for each step of the operations. Unlike GMP facilities, assigning separate rooms for each task is not mandatory for GTP. However, the workflow along the different subdivisions of the facility should be clear to prevent contamination or cross-contamination.

Suitable lighting, ventilation, and plumbing are essential. Careful layout should take into consideration the practical easy access to sinks and toilets.

7.2.1.1.1 Good tissue practice environmental control

Environmental control of GTP facilities includes proper control of the operation conditions, which are temperature and humidity, ventilation and air filtration, cleaning and disinfection of rooms and equipment, as well as regular maintenance of the equipment.

Contrary to GMP facilities, no clean room classifications are required in GTP. Environmental control levels are left to the protocol specifications.

7.2.1.2 Good manufacturing practice

GMP refers to the combined specifications that ensure consistent quality and safety of the end-products. GMP's implementation involves what is known as "5P": premises, processes, procedures, people, and products.

7.2.1.2.1 Facilities

GMP facilities are tightly controlled premises known as clean room facilities.

Clean room facilities have the following characteristics:

1. Epoxy (vinyl) floors
2. Smooth floor—wall transition
3. Cleanable walls and doors
4. Nonshedding and nonporous walls
5. Washable paint and coatings
6. Impermeable bench tops
7. Interlock doors

Each clean room has a total area of approximately 500 m^2. This area includes the lab rooms and service corridors. It is divided into classes according to the allowed particle count. The controlled environment of the clean room is of utmost importance. Environmental control includes air circulation and quality, pressure, temperature, and humidity.

Air: Clean room premises are divided into two parts: outer part (half of the area), with particle count of less than 100,000 per cubic foot or air, and the inner part with no more than 100 particles per cubic foot of air. Air change in the clean room area should be designed as 20—30 times per hour with unidirectional air flow from ceiling to floor. All air entering the clean room should pass by HEPA (high-efficiency particulate air) filters.

Pressure: Clean rooms should have a positive pressure to reduce the entry of any contaminants.

Temperature: Temperature should be stable (no more than 1°C change) with constant humidity at 40% with a range of no more than 2% variability.

Classification:

According to allowed particle count, clean rooms are classified into:

Class D "unclassified"

Class C

Class B

Class A

These areas are arranged in descending order of allowed particle count, with Class D being the highest allowed particle count and class A with lowest allowed particle count. The flow of work in the clean room goes as follows: *Class D* area is the first part of the stem cell center where administrative and clerical functions are implemented, with the offices and computers needed for recording data and patients and/or sample data. This area should have the bar coding instruments for sample receipt.

Class C laboratory contains the nonsterile laboratory equipment for routine testing, such as cell count, viability, or molecular identification of samples not to be injected or further cultured. *Class B* is the sterile laboratory area where cell separation and identification equipment and procedures are performed. *Class A* is the laminar airflow sterile workstation inside Class B laboratory.

Gowning areas are located between C and D, so that entering the laboratory workplace is preceded by gowning, and between C and B, so that entering the sterile laboratory requires further gowning (Table 7.2).

Handling of cellular products is performed in class A laboratory (Fig. 7–2).

7.2.1.2.2 Equipment

Equipment requirements are (Table 7.3):

1. Equipment must be of suitable design.
2. Appropriate location and installation that ensures the easy cleaning and disinfection as well as prevention of the introduction or transmission of contamination.
3. Equipment must be periodically calibrated to ensure valid results.
4. Equipment should be sanitized before and after dealing with tissues.

7.2.1.2.3 Basic equipment for stem cell laboratory are

1. Tissue culture laboratory equipment:
   a. Class II biosafety cabinet
   b. CO_2 incubator
   c. Clinical grade low speed centrifuge
   d. Water bath
   e. Pipettors
2. Stem cell separation equipment:
   a. Immunomagnetic separation, clinical grade
   b. Sepax
   c. Adipose:

Table 7.2 Classification of clean room areas.

	Class D	Class C	Class B	Class A
Purpose	Administrative	Quality control testing of sample	1. Cell wash 2. Microscopic Identification of cells 3. Cell count	Separation, culture, and expansion of stem cells
Equipment	Data, bar coding	1. Flow cytometer 2. Molecular Biology (Polymerase chain reaction [PCR] and sequencing unit 3. Ordinary microscopes 4. Phase-contrast microscope	1. Laminar airflow cabinet 2. CO_2 incubator 3. Centrifuge 4. Inverted microscope	Laminar airflow cabinet
Personnel	Administrative	Routine laboratory	Cell culture	Cell culture
Gowning	None	First gowning	Second gowning	Second Gowning
Particle count	100,000	100,000	100	100
Type of samples handled	First receipt of samples	Samples for testing (not re-injectable)	Sterile injectable samples (closed system)	Sterile injectable samples

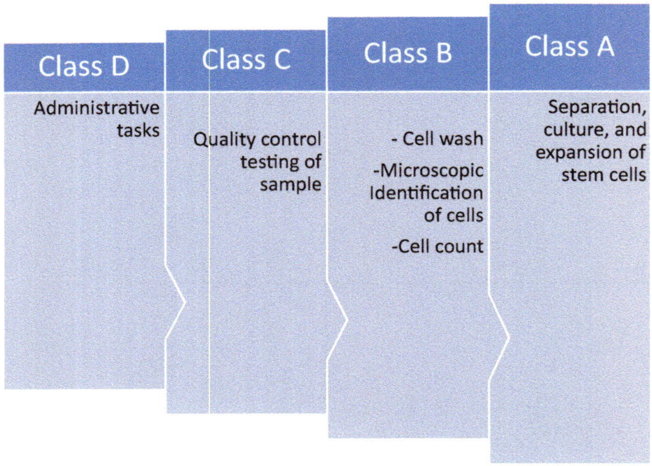

Figure 7–2 Workflow in a clean room.

   d. Hemopoietic
   e. MSC
3. Microscopy:
   a. Inverted microscope
   b. Phase-contrast microscope
   c. Ordinary microscope
4. Storage equipment:
   a. Refrigerator (4°C)
   b. Freezer −20°C
   c. Low temperature freezer (−70°C to −85°C)
   d. Cryogenic freezer (−180°C)
5. Quality control equipment:
   a. Molecular biology: RT-PCR
   b. Flowcytometry
   c. Fluorescent microscope
   d. Confocal microscope
6. Core facility:
   a. Microarray gene expression
   b. Genomics
   c. Proteomics
7. Other facilities:
   a. Back up electricity generators
   b. CO_2 and liquid nitrogen tank or cylinders delivery areas

Table 7.3 Equipment requirement for stem cell laboratory.

Name	Function	Site	Classification
1. Laminar airflow cabinet	Sterile site for stem cell handling: separation, culture, and passage	Tissue culture laboratory	Class A
2. Cooling centrifuge	Physical stem cell separation and washing steps during stem cell handling	Tissue culture laboratory	Class B
3. Inverted microscope	Morphological evaluation of stem cell cultures	Tissue culture laboratory	Class B
4. Water bath	Temperature adjustment of reagents before use, incubation steps during stem cell processing	Tissue culture laboratory	Class B
5. Stem cell separation instruments (Clinimax, Sepax, etc.)	Closed system stem cell separation	Tissue culture laboratory	Class B
6. Ordinary microscope	Evaluation of stem cell viability, cell count	Routine laboratory	Class B
7. Flow cytometer	Stem cell identification, characterization, evaluation of apoptosis	Routine lab	Class C
8. PCR system	Evaluation of gene expression in effectively manipulated stem cells	Routine lab	Class C
9. Refrigerators: −4°C, −20°C, −80°C, and cryogenic freezers	Stem cell and media storage	Routine lab	Class C
10. Microarray	Evaluation of gene expression in effectively manipulated stem cells	Core lab	Unclassified

c. Storage area for laboratory disposables
d. IT areas

7.2.1.2.4 Supplies and reagents

All reagents used for processing of stem cells must be clinical grade. Sterile supplies, such as tubes, pipettes, and flasks, must meet the specifications designed to reduce the risk of contamination.

7.2.1.2.5 Processing and process control

Stem cell handling system:

SOPs must be established for all major and minor procedures in a stem cell therapy facility. This includes all steps of the process, starting from sampling or sample receipt, coding, stem cell processing, stem cell validation and identification, sterility check, and preparation for injection to final delivery of the sample. This applies for the laboratory part of the procedure.

Regarding the clinical part of the procedure, SOPs for clinical assessment of the patient, informed consent, stem cell injection procedure, and evaluation must be prepared.

Detailed SOPs for all steps of the stem cell therapy procedure are included in the APPENDIX.

7.2.2 Personnel qualifications and job description

7.2.2.1 Administrative team

The role of the administrative team is to adjust the workflow of stem cell processing and characterization. Steps start from sample receipt, recording, and coding. Setting the timetable of stem cell separation, culture, passage, and harvest, as well as coordination between subunits of the facility is one of the important duties of the administration.

The next step is scheduling and recording of viability, sterility checks, as well as genetic stability testing. Lastly, the cells are carefully stored.

Therefore, the main duties of the administrative team are preparing worklists for the different departments according to the SOPs and work teams according to the scheduled tasks and acting as a liaison between different departments.

7.2.2.2 Ethics team

The responsibilities of the ethics team start from reviewing the compliance of the procedures and SOPs with the regional governing ethical standards.

This is followed by composing the consent in a way that reflects the procedures. This job requires thorough knowledge of medical ethics and a close follow up of guidelines and regulations.

7.2.2.3 Clinical evaluation team

The clinical evaluation team is responsible for diagnosis of problem, type of cellular defect, and lineage of the cells to be used. In addition, the clinical team is responsible for choosing the route of administration as per their knowledge of the physiology of the target organ. Choice of the evaluation method for stem cell therapy is also allocated to the clinical team as related to the presenting symptom of organ failure. To undertake this job properly, the clinician should have extensive knowledge of the medical discipline involved (neurology, cardiology, orthopedics, dermatology, etc.). In addition, the clinician should be familiar with the basics of regenerative medicine.

7.2.2.4 Stem cell collection team

Stem cell collection varies according to the type and source of stem cells. Clinically applicable stem cell types are adult hemopoietic stem cells (HSCs), cord blood HSCs, adult MSCs, cord blood and tissue MSCs, and adipose-derived stem cells. Detailed SOPs of these procedures can be referred to in the APPENDIX. These different procedures require various skills and experience in the relevant procedure. The most important qualifications or skills required in these teams include thorough experience with sterile procedures and sterile tissue handling.

7.2.2.5 Stem cell handling team (laboratory)

Stem cell handling team is the team dealing with the stem cell isolation, culture, expansion, and characterization. This is the core process in the whole facility. Members of this team should have extensive knowledge of sterile tissue culture techniques, stem cell isolation, culture, and characterization procedures, as well as meticulous recording of the whole experiment.

7.2.2.6 Clinical followup team

The clinical followup team is responsible for the periodic evaluation of the patients after injection using the predetermined clinical grading or scoring system (APPENDIX). This team makes use of the laboratory stem cell tracking data.

7.3 Quality control team

The quality control team is responsible for the following:

1. Evaluating the matching of the actual procedures to the acknowledged SOPs.
2. Analyzing the stem cell characterization and sterility check data.
3. Keeping record of the personnel job descriptions, continuous education, and training schedule.
4. Facility process organization, reporting structures, and hierarchy
5. Material supply management and supplier qualifications
6. Equipment calibration and maintenance
7. Managing nonconforming events

The quality control personnel should have, in addition to the quality control qualification, basic knowledge of the procedures and the overall process in stem cell therapy.

Further reading

Basford C, Forraz N, Habibollah S, Hanger K, McGuckin C. The cord blood separation league table: a comparison of the major clinical grade harvesting techniqes for cord blood stem cells. Int J Stem Cell 2010;3(1):32–45.

Bennett S, Johnson N, Lasky L. Quality assurance and standards in hematopoietic progenitor processing. J Clin Apher 1992;7(3):138–44.

Daniel N, Zinno F. Quality control of cryopreserved hematopoeitic stem cells. Acta Hematol Pol 2015;46(5):347–52.

George B. Regulations and guidelines governing stem cell based products: clinical considerations. Perspect Clin Res 2011;2(3):94–9.

Guidance for Industry: Current Good Tissue Practice (cGTP) and Additional Requirements for Manufacturers of Human Cells, Tissues and cellular and tissue-based products (HCT/Ps): U.S. Department of Health and Human Services. Food and Drug Administration. Center for Biologics Evaluation and Research, 2011.

Guidelines for stem cell research and clinical translation, 2016. http://www.isscr.org.

Kin T, Mirbolooki M, Lakey J. Design and implementation of a cell processing facility. Cellular Transplantation: From Laboratory to Clinic. Academic Press; 2007. p. 629–33.

Lee T, Lysaght T, Lipworth W, Hendl T, Kerridge I, Munsie M, et al. Regulating the stem cell industry: needs and responsibilities. Bull World Health Organ 2017;95:663–4.

Lee T, Lysaght T, Lipworth W, Hendl T, Kerridge I, Munsie M, et al. Regulating the stem cell industry: needs and responsibilities. Bull World Health Organ 2017;95:663–4.

Leemhuis T, Padley D, Keever-Taylor C, Niedervieser D, Teshima T, Lanza F, et al. Essential requirements for setting up a stem cell processing laboratory. Bone Marrow Transplant 2014;49:1098–105.

Lomax G, Torres A, Millan M. Regulated, reliable, and reputable: Protect patients with uniform standards for stem cell treatments. Stem Cell Transl Med 2020;9(5):547–53.

Mijovic A, Pamphilon D. Harvesting, processing and inventory management of peripheral blood stem cells. Asian J Transfus Sci 2007;1(1):16–23.

National Guidelines for Stem Cell Research. Indian Council of Medical Research & Department of Biotechnology 2017. Published by the Division of Publication and Information on behalf of the Secretary DHR & DG, ICMR, New Delhi. Printed at M/s Royal Offset Printers, A-89/1, Naraina Industrial Area, Phase-I, New Delhi-110028.

Nicodemou A, Danisovic L. Mesenchymal stromal/stem cell separation methods: concise review. Cell ad Tissue Bank 2017;18:443–60.

O'Shea O, Steeg R, Chapman C, Mackintosh P, Stace N. Development and implementation of large-scale quality control for the european bank for induced pluripotent stem cells. Stem Cell Res 2020;45:101773.

Shparberg R, Braunsteiner N, Russell Vickers E. Laboratory training guidelines for clinicians undertaking stem cell therapy. J Stem Cell Res Ther 2018;8(5):427.

Regenerative and Stem Cell Therapy Practices Report and Recommendations of the Workgroup to Study Regenerative and Stem Cell Therapy Practices Adopted as policy by the Federation of State Medical Boards April 2018.

Reisman M, Adams K. Stem cell therapy: a look at current research, regulations, and remaining hurdles. P T 2014;39(12):846–7 854–857.

Rodriguez J, Pratta AS, Abbassi N, Fabre H, Rodriguez F, Debard C, et al. Evaluation of three devices for the isolation of the stromal vascular fraction from adipose tissue and for ASC culture: a comparative study. Stem Cell Int 2017; Article ID 9289213, 14 pages.

Sekiya E, Forte A, Kuhn T, Janz F, Bydlowski S, Alves A. Establishing a stem cell culture laboratory for clinical trials. Rev Bras Hematol Hemoter 2012;34(3):236–41.

Watts M, Linch D. Optimisation and quality control of cell processing for autologous stem cell transplantation. Br J Hematol 2016;175:5.

Wesselschmidt R, Schwartz P. The stem cell laboratory: design, equipment, and oversight. Methods Mol Biol (Clifton, NJ) 2011;767:3–13. Available from: https://doi.org/10.1007/978-1-61779-201-4_1.

CHAPTER 8

Protocol guidelines for clinical decisions in stem cell therapy

Contents

8.1 Introduction 99
 8.1.1 Timing of intervention 99
 8.1.2 Stem cell injection 99
 8.1.3 Frequency of injection 101
Reference 103
Further reading 103

8.1 Introduction

As outlined earlier, thorough clinical examination and investigations are mandatory for an accurate diagnosis. Clinical details of the treatment regimen, starting from timing of treatment, route of injection, number of injections until choosing the method for evaluation, are dependent on the diagnosis. In this chapter, we will outline the most important clinical decisions regarding stem cell therapy.

8.1.1 Timing of intervention

Stem cell injection can be done in the acute, subacute, or chronic phase of the disease. The advantages and disadvantages of each intervention is outlined in Table 8.1.

8.1.2 Stem cell injection

8.1.2.1 Route of injection

The optimum route of stem cell injection is subject to intensive research. Numerous studies have compared the effect of the route of the injection of stem cells on their therapeutic effects. In some disciplines, studies favored a certain route, while in others, results are still controversial. The

Stem Cell Therapy
DOI: https://doi.org/10.1016/B978-0-12-821569-2.00010-7

Table 8.1 Comparison between the timing of stem cell injection.

	Acute lesion	Subacute lesion	Chronic lesion
Advantages	1. Efficient homing of stem cells due to inflammatory cytokines 2. No fibrosis 3. Optimal blood supply	1. Fair homing of stem cells due to the residual inflammatory cytokines 2. No fibrosis 3. Fair blood supply	1. Good engraftment of the cells due to the absence of inflammatory cytokines 2. Transdifferentiation into lineage cells due to the dominant tissue microenvironment
Disadvantages	1. Possible apoptosis in injected stem cells due to the hostile microenvironment 2. Injected stem cells can differentiate into macrophages instead of lineage cells	1. Less engraftment due to the remaining inflammatory cytokines	1. Fibrosis provides poor microenvironment and blood supply to injected cells 2. Poor homing due to the absence of inflammatory cytokines

choice of the injection modality depends on a number of variables, for example:

1. *Variables related to the organ*:
   This is perhaps the most important variable influencing the route of injection. Each organ has special physiological and metabolic characters that favor one or the other routes of injection. The type of blood supply, the physiological functioning unit of the organ, and the location of the stem niche are all determinants of the route of injection.
2. *Variables related to the type of injury*:
   a. Type of cellular injury
   b. Vascular condition of the organ
   c. Presence or absence of fibrosis.
3. *Variables related to the type of stem cell used*:
   Using undifferentiated stem cells aiming to achieve transdifferentiation induced by microenvironmental factors, the injection should be local, while differentiated cells can be injected in the blood supply.

According to the proposed mechanism of action of the stem cells, cells are injected in the following sites (Table 8.2):

a. Arterial: organs with double blood supply, for example, the liver, the lungs, the testes, and the ovaries
b. Venous
c. Parenchymal
d. Subcapsular.

8.1.3 Frequency of injection

Injected stem cells need healthy microenvironment, adequate blood supply, and supportive niche to engraft and propagate. These tissue resources can only accommodate a finite number of cells. Engrafted cells will further contribute to these tissue resources, enabling the tissue to accommodate another number of stem cells. Thus, dividing the therapeutic stem cell dose over a number of injections will lead to a higher number of engrafted stem cells than if injected all as a single injection.

This applies to healthy organs and is further augmented in organs with a limited blood supply or unhealthy microenvironment due to fibrosis or infarction.

As paracrine effects of mesenchymal stem cells (MSCs) are the most important mode of regeneration, the period of survival of MSCs will greatly affect their therapeutic potential. Ref. [1] demonstrated longer survival of MSCs following double low dose of MSCs than a single high dose. Thus, the injection frequency is a crucial factor in MSC therapy.

8.1.3.1 Methods of evaluation

Evaluation of stem cell therapy regarding efficacy and the incidence of any complications is performed on two levels: clinical evaluation of the patient and following the cell trafficking and physiology inside the body.

8.1.3.1.1 Patient

1. Clinical assessment of the patient is performed after a careful history taking. Listening to the patient's story of any change in the quality of life or disease state or progression is crucial. Afterward, clinical examination is performed to verify such findings.
2. Assessment scales: they are semiquantitative method of evaluation of the change in the clinical state of the patient. They are essential to

Table 8.2 Comparison between the routes of stem cell injection.

	Vascular		Parenchymal	Subcapsular	Systemic
	Arterial	Venous			
Advantages	Direct Homogenous cell engraftment	Longer cell retention	Direct Accurate		Simple Noninvasive
Disadvantages	Cells do not reach unperfused or fibrotic areas	Individual variability in venous anatomy	Invasive		Low cell delivery Used only in acute conditions (homing)

eliminate evaluator bias and to provide an objective evaluation of the case. Two types of assessment scales are used in this setting:

a. Clinical evaluation scales: such as American spinal injury associaton scale in spinal cord injury.

b. Quality of life scale: Disease-oriented quality of life scales are used to verify the overall effect of the therapy on a patient's life.

8.1.3.1.2 Cells

Cell tracking in the clinical setting is extremely difficult. Iron oxide tagging is the only clinically approved method of cellular tracking. Iron oxide lablelled cells can be detected either by tissue biopsy or MRI imaging.

Reference

[1] Kang WC, Oh PC, Lee K, Ahn T, Byun K. Increasing injection frequency enhances the survival of injected bone marrow derived mesenchymal stem cells in a critical limb ischemia animal model. Korean J Physiol Pharmacol 2016;20(6):657–67.

Further reading

Calvin CS, Li Z, Jijun H. Current stem cell delivery methods for myocardial repair. BioMed Res Int 2013;2013. Available from: https://doi.org/10.1155/2013/547902 Article ID 547902, 15 p.

Chang H, Bhatt S, Dogra V. Pearls and pitfalls in diagnosis of ovarian torsion. Radiographics 2008. Available from: https://doi.org/10.1148/rg.285075130.

Gounder SS, Kodiappan R, Chan SC, Veerakumarasivam A, Subramani B. Adult stem cell therapy in liver cirrhosis management: current practices and future perspectives. Int J Stem Cell Res Ther 2017;3:049. Available from: https://doi.org/10.23937/2469-570X/1410049.

Idriss NK, Sayyed HG, Osama A, Sabry D. Treatment efficiency of different routes of bone marrow-derived mesenchymal stem cell injection in rat liver fibrosis model. Cell Physiol Biochem 2018;48:2161–71.

Irodi A, Cherian R, Keshava SN, James P. Dual arterial supply to normal lung: within the sequestration spectrum. Br J Radiol. 2010;83(989):e86–9. Available from: https://doi.org/10.1259/bjr/30458107.

Kurtz A. Mesenchymal stem cell delivery routes and fate. Int J stem Cell 2008;1(1):1–7. Available from: https://doi.org/10.15283/ijsc.2008.1.1.1.

Mostafa T, Labib I, El-Khayat Y, El-Rahman El-Shahat A, Gadallah A. Human testicular arterial supply: gross anatomy, corrosion cast, and radiologic study. Fertil Steril 2008; 90(6):2226–30. Available from: https://doi.org/10.1016/j.fertnstert.2007.10.013 Epub 2008 Jun 13. PMID: 18555239.

Mostafa T, Labib I, El-Khayat Y, El-Rahman El-Shahat A, Gadallah A. Human testicular arterial supply: gross anatomy, corrosion cast, and radiologic study. Fertil Steril 2008;90(6):2226–30. Available from: https://doi.org/10.1016/j.fertnstert.2007.10.013 Epub 2008 Jun 13. PMID: 18555239.

Pan K, Deng L, Chen P, Peng Q, Pan J, Wu Y, et al. Safety and feasibility of repeated intrathecal allogeneic bone marrow-derived mesenchymal stromal cells in patients with neurological diseases. Stem Cell Int 2019;2019:8421281.

Plock J, Schnider J, Schweizer R, Zang W, Tsuji W, Waldner M, et al. The influence of timing and frequency of adipose-derived mesenchymal stem cell therapy on immunomodulation outcomes after vascularized composite allotransplantation. Transplantation 2016;101(1):1.

Zhou GP, Jiang YZ, Sun LY, et al. Therapeutic effect and safety of stem cell therapy for chronic liver disease: a systematic review and meta-analysis of randomized controlled trials. Stem Cell Res Ther 2020;11:419. Available from: https://doi.org/10.1186/s13287-020-01935-w.

CHAPTER 9

Protocol guidelines: laboratory decisions in stem cell therapy

Contents

9.1 Introduction 105
9.2 Choosing the cell purification level 106
9.3 Choice of purification strategy 106
9.4 Choice of stem cell differentiation status 107
9.5 Choice of cell dose 108
9.6 Choice of tracking method 109
9.6.1 Methods of in vivo cell tracking 109

9.1 Introduction

Stem cell therapy is a multidisciplinary field of medicine that comes at the intersection of basic science, clinical practices, and translational research. In the previous chapter, we outlined the regulatory framework for the facilities offering stem cell manipulation and therapy. In this chapter, we will discuss the regulations governing the laboratory procedures involved in stem cell preparation.

Laboratory procedures involved in stem cell preparation can be divided into stem cell isolation, stem cell culture or expansion, induction of transdifferentiation, stem cell tagging, verification of sterility and genetic integrity, and finally, preparation for injection (Fig. 9−1).

A number of questions must be answered to finalize the laboratory protocol for preparation of stem cells for regenerative solutions. Choosing the optimum lineage, source, cell purification, cell differentiation, dose, and cell tracking are the important questions that need to be answered. In the previous chapter, we discussed the differences between different stem cell lineages and sources in transdifferentiation potential. In this chapter, we will discuss important laboratory decisions such as purification, expansion, differentiation, dose, and tracking.

Stem Cell Therapy
DOI: https://doi.org/10.1016/B978-0-12-821569-2.00009-0

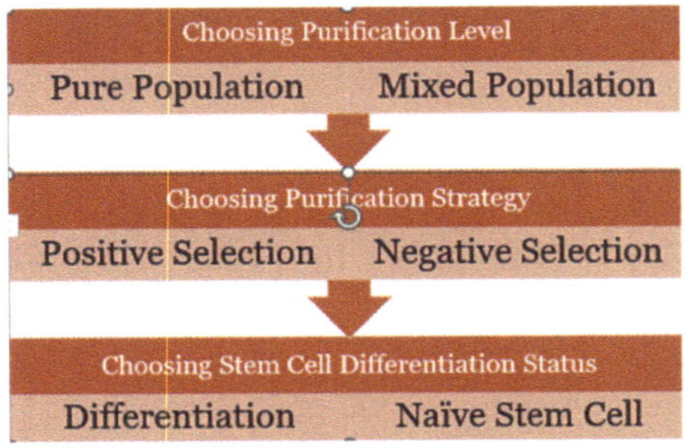

Figure 9–1 Process of strategy development for stem cell therapy.

9.2 Choosing the cell purification level

Stem cell isolation is usually a multistep procedure, which starts with the isolation of a mixed population and proceeds to purification of the target cell population. The decision to purify stem cells before injection is subject to the following parameters:

1. Site of injection: the capacity of the tissue to accommodate the number of cells and the volume of the fluid in which the cells are suspended varies among body tissues. Tight sites, such as the joints, eye, and testis, do not accommodate a large volume; therefore, the cells have to be suspended in minimal amount of fluid. On the other hand, loose tissues, such as the liver, can accommodate a larger fluid volume. The smaller the amount of fluid, the lower the total number of cells injected. Thus, this mandates purification of the cell population to obtain the target number of stem cells.
2. Immunological "privileged" sites: some tissues in the human body are protected from the immune competent cells, for example, testis and eye. The antigens of these tissues are not detected as "self-antigens" by the body's immune cells. Delivering mixed cell population to these tissues in cell therapy carries the risk of initiation of immune reaction against these tissues. Thus, purification of the cellular therapy product is required to avoid initiation of such process.

9.3 Choice of purification strategy

Planning of stem cell purification strategy depends on the type of the cell needed. Stem cell selection can either follow *positive selection* or *negative selection*. Positive selection is a procedure where the stem cell is targeted directly, leaving the unwanted cells in the mixture, while in negative selection, other cell populations are targeted and eliminated, leaving the stem cells in the mixture (Table 9.1).

9.4 Choice of stem cell differentiation status

Stem cell therapy can use undifferentiated stem cells or differentiated stem cells. As discussed previously, the modes of regenerative function of stem cells include transdifferentiation as well as modulation of microenvironment in the form of induction of angiogenesis, antiinflammatory, immunomodulation, or stimulation of endogenous stem cells. Thus, the role of transdifferentiation in the induction of regeneration is not as important as previously thought. Transdifferentiation can occur spontaneously after stem cell injection mediated by the microenvironmental factors, such as growth factors and proteins. Ex vivo induction of transdifferentiation can also be used before stem cell injection.

The choice of injection of noninduced or induced stem cells is important in planning stem cell therapy (Table 9.2).

Thus, using noninduced stem cell population provides a more viable cell population with a longer lifespan and better incorporation and engraftment within the target tissue. However, the functional replacement

Table 9.1 Comparison between positive and negative stem cell selection approaches.

	Positive selection	Negative selection
1. Targeted population	Stem cell population	Nonstem cell populations
2. Indications	1. Well-defined stem cell population with specific surface markers 2. High purity levels required 3. The surface receptors used are not required in downstream experiments	1. The stem cell population is not well defined 2. A "nonstem population" must be eliminated, for example, T-lymphocytes In immune disturbances
3. Disadvantages	Stem cell manipulation may affect viability	High cost due to the large number of antibodies used

Table 9.2 Comparison between injection of induced and noninduced stem cells.

	Induced	Noninduced
1. Engraftment	Low	High
2. Function	Immediate	Delayed
3. Viability	Reduced due to manipulation	Better (less manipulation)
4. Lifespan	Limited	Unlimited (self-renewal)
5. Mode of regenerative function	Replacement of cell population	1. Induction of angiogenesis 2. Modulation of microenvironment 3. Stimulation of endogenous stem cells 4. Replacement of cell population (transdifferentiation)
Disadvantages	1. High cost 2. Cell manipulation carries the risk of compromised viability or sterility 3. Nonincorporation within the tissue	Delayed effect of regeneration

may be delayed until the transdifferentiation or maturation of endogenous stem cells occurs.

A notable disadvantage of using induced or transdifferentiated stem cells is that cells start functioning before complete harmonization with the receptive tissues. This can lead to arrhythmia in cardiac tissue or aberrant neural impulses and connections in the nervous tissue.

Therefore, the standard practice is to use undifferentiated stem cells unless immediate replacement of function is required.

9.5 Choice of cell dose

Determining the appropriate cell dose follows many parameters as discussed in the previous chapter. In this chapter, we will outline the choice of ex vivo expansion of stem cells before injection. Achieving the target cell dose is of great importance to the success of stem cell therapy. Sometimes, however, the required dose cannot be isolated from the chosen source. Ex vivo expansion can be used to achieve it. However, ex-vivo expansion of the stem cell population can induce premature senescence of the injected cells as each passage of the cells reduces stem cell self-renewal. Passaging of stem cells can also reduce the ability of the cells to differentiate.

Another important disadvantage of ex vivo expansion is reduced genetic stability. The incidence of DNA damage in the cultured cells increases with passaging.

Balancing the optimum cell dose with the best viability is a delicate decision.

9.6 Choice of tracking method

Tracking of injected stem cell population is essential for safety and delineating the mechanism of action of stem cells.

9.6.1 Methods of in vivo cell tracking

9.6.1.1 Experimental

In experimental animals, cells can be tracked through gene tracking. The marker gene may be:

1. X/Y chromosomes in gender-mismatched transplantations
2. Transgenic animals
3. Green fluorescent protein (GFP)-marked cells

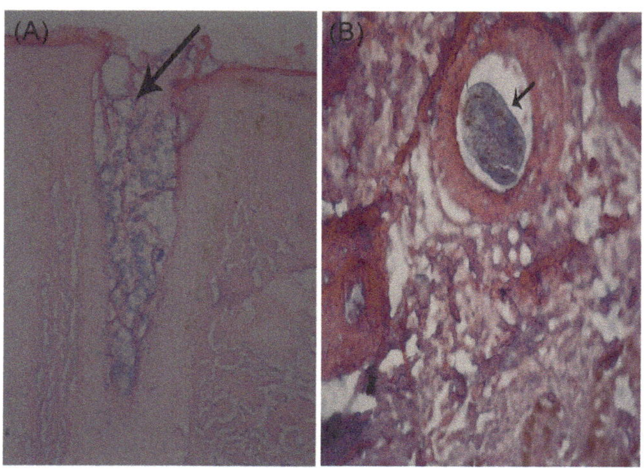

Figure 9–2 Iron oxide labeling of hair follicle stem cell in a case of stem cell therapy for alopecia. The figure shows iron oxide positivity all through the hair follicle in longitudinal section (A) and transverse section (B).

9.6.1.2 Methods of in vivo tracking in humans

1. Gender-mismatched transplantations: gender-mismatched cells are tracked using XY probes in tissue biopsies.
2. Iron oxide cell labeling (Fig. 9–2).

Characteristics of an ideal imaging technology for stem cell tracking during clinical trials:

1. Biocompatible, safe, and nontoxic
2. No genetic modification or perturbation to the stem cell
3. Cell detection at any anatomic location
4. Quantification of cell number
5. Minimal or no dilution with cell division
6. Minimal or no transfer of contrast agent to the cells
7. Noninvasive imaging of the living subject over months to years
8. No contrast agent required for injection

CHAPTER 10

Phases of translational stem cell research

Contents

10.1 Introduction 111
10.2 Laboratory experiments 111
10.3 Preclinical, experimental 112
  10.3.1 Value of animal experiments 112
  10.3.2 Small versus large animals 116
  10.3.3 Challenges in animal experiments 117
  10.3.4 Ethical considerations 117
10.4 Clinical trials 117
Further reading 118

10.1 Introduction

In order to move the stem cell therapy idea from concept to benchtop to bedside, three important steps are required. First, laboratory experiments are required to validate the type of cells required and the optimum source, as well as for standardization of laboratory procedures. The second step involves the application of this cellular therapy procedure in an animal model of the target disorder. The importance of this experimental animal procedure is to optimize the cell dose, route, and clinical efficacy.

The last step is a clinical trial in humans. This is divided into the known phases: a pilot clinical trial and phases I, II, III, and IV.

10.2 Laboratory experiments

Laboratory experiments are the starting point of application of a "regenerative stem cell therapy" idea. As mentioned previously, laboratory experiments are designed to answer fundamental questions, such as the choice of cell lineage, the source, and differentiation. Comparative laboratory experiments should be conducted to finalize protocols for effective

Stem Cell Therapy
DOI: https://doi.org/10.1016/B978-0-12-821569-2.00008-9

isolation, purification, and transdifferentiation. Finally, laboratory testing for sterility and genetic stability are required after the last steps of laboratory protocol to release the cells for use.

10.3 Preclinical, experimental

After completing in vitro experiments, evaluating the selected cells "in action" is mandatory. The integration of the cells in living tissues and their engraftment and function is subject to numerous variables that can affect the overall regenerative value of the cells. The interaction between the injected cells, resident cells, and extracellular matrix is complex and can never be precisely mimicked in the laboratory. Possible regenerative mechanisms and adverse cell–cell and cell–matrix interactions can only be examined in live animal experiments.

10.3.1 Value of animal experiments

1. Cell dose: determination of the optimum effective cell dose can be done in dose escalation studies in animal experiments. Information of animal studies is not exactly mimicked in human trials; however, it can serve as a guide for pilot clinical trials. Dose calibration studies include the number of cells, the volume of the cell suspension, and the number and schedule of injections.
2. Route of administration: comparing different routes of administration of stem cell therapy is one of the important targets of preclinical animal studies. Evaluation of any difference in clinical efficacy, complications, or cell fate is noted.
3. Cell tracking and function: the life cycle of cells after injection, their incorporation into the target tissue, and any atypical cellular reaction surrounding the cells are studied in histological sections. Cell tracking to follow circulation and recirculation in the body should be done. Evaluation of cell function after injection is done according to the cell lineage involved.
4. Therapeutic efficacy: therapeutic efficacy of the injected stem cells is done through clinical examinations and laboratory tests. Clinical examination should be as objective as possible, applying quantitative parameters and scales. Laboratory tests are used to evaluate functional parameters of the organs as well as markers of inflammation.
5. Mode of regenerative function: animal studies are important in studying the mode of regenerative function using techniques that cannot be

used in the human trials. Tissue biopsy, tracking using fluorescent tags, and autopsy are used to study the effect of cells on surrounding tissues as well as the whole body.

10.3.1.1 Animal models

Animal modeling and treatment are pivotal steps in translational research and act as an informative bridge between laboratory experiments and human application. This kind of modeling serves in planning optimum protocols for human trials and reducing complications to human beings.

Animal models for human diseases are either spontaneous or induced. Induction of animal models varies according to the pathogenesis of the disease. In unifactorial diseases, for example, vascular or traumatic diseases, modeling is straightforward and easy. However, in multifactorial diseases, it is more complex. When the disease is caused by infectious agents, we have to be extremely careful as infectious agents cause variable pathologies in different species. Thus, choosing the most influential pathology in case of multifactorial or infectious diseases is an important choice in creating the animal model. For example, in viral liver cirrhosis, the important pathology is periportal fibrosis. Thus, although human hepatitis virus does not cause cirrhosis in mice, inducing periportal fibrosis using chemical agents will produce an animal model close to the pathology of human liver affected by hepatitis virus.

10.3.1.2 Species selection

Another important parameter to be kept in mind when working with animal models is the difference in anatomical structure of the target organ, for example, the different ratio between white and gray matter in human versus rodent brain. This must be kept in mind when interpreting results of experiments.

In translational stem cell research, the choice of the animal species for the model is also governed by the availability of imaging and molecular techniques employed for this species. Thus, despite the large number of animals available, only a few can be conveniently used for modeling human disorders.

10.3.1.3 Strain selection

Definite biological differences are observed between different strains of animals. This can lead to variations in experimental results according to the species used. The idea of "inbred" strains was introduced to minimize these differences.

10.3.1.4 Methods of inducing animal models

1. Surgical: Induction of animal models of traumatic and vascular diseases can be done through surgical procedures, such as ligation of vessels for induction of infarction or stroke, cutting of nerve or spinal cord to induce paralysis, or excision of organs or parts of organs, such as partial hepatectomy or nephrectomy.
2. Chemical: Chemicals can be used to induce tissue damage mimicking the pathological features of certain human diseases. Numerous examples are present in literature: using carbon tetrachloride for inducing liver fibrosis, using ethidium bromide to induce encephalomyelitis, and induction of inflammatory bowel disease using trinitrobenzene sulfonic acid, dextran sodium sulfate, or oxazolone. A large number of chemicals have been successfully used to induce diseases such as diabetes, liver fibrosis, pulmonary fibrosis, pulmonary hypertension, and parkinsonism, and among others (Table 10.1).
3. Biological: Induction of animal models using biological mechanisms involve using either pathogens (bacterial, viral, or parasitic) or cellular injection, for example, injection of allogenic bone marrow to induce graft versus host disease (Table 10.2).
4. Gene manipulation: Gene manipulation to produce animal models of disease has been extensively used in clinical research. This can use "transgenic" technology in which foreign DNA is introduced into the

Table 10.1 Chemicals used for induction of animal models of human diseases.

	Chemical	Disease
Dermatological disorders	5% imiquimod	Psoriasis
Autoimmune disorders	Collagen in complete Freund's adjuvantMyelin oligodendrocyte glycoprotein in complete Freund's adjuvant	Rheumatoid arthritisMultiple sclerosis
Neurological disorders	Heavy metalsEthanol	Alzheimer's diseaseParkinson's disease
	Streptozotocin	Diabetes
Pulmonary diseases	Bleomycin	Pulmonary fibrosisPulmonary hypertension
Cancer	Benzene	Leukemia

Table 10.2 Methods of induction of animal models.

Chemical	Surgical	Genetic	Immune	Infectious
1. Induction of oncogenesis: (e.g., N-nitroso compounds) 2. Induction of liver fibrosis: using carbon tetracholride 3. Induction of diabetes using streptozotocin 4. Induction of neural degeneration using ethedium bromide 5. Induction of Parkinsonism using 1-methyl-4-phenyl-1,2,3, 6-tetrahydropyridine (MPTP) or 6-hydroxydopamine (6-OHDA)	1. Induction of stroke by occlusion of middle cerebral artery in rats. 2. Induction of acute renal failure by occlusion of renal artery 3. Induction of osteoporosis by bilateral ovariectomy	1. Single gene knock-out and knock-in 2. Conditional gene modification 3. Transgenesis	Immunization with auto-antigens to induce auto immune disease (e.g., experimental autoimmune encephalomyelitis)	1. Injection of pathogens to induce infections 2. Injection of pathogens or toxins to induce sepsis

embryo or after birth, or it can use postnatal approach. DNA manipulation can introduce a specific mutation "knock-in" or cause the loss of a certain gene function "knock-out."

10.3.2 Small versus large animals

Small animals such as mice, rodents, and rabbits are easy to breed and are cheaper in maintenance. They can be adequate for most functional experiments. However, large animals such as canine and chimpanzees are preferred in musculoskeletal and cognitive experiments (Table 10.3).

Table 10.3 Comparison between small and large animals in disease models.

	Small Animals (Rodents)	Large animals (Non-rodent mammalian species)
Advantages	1. Rodents can be easily modified genetically to reproduce human disease. 2. Low cost 3. Rapid reproductive cycle 4. Easier ethical standards	1. Some diseases can occur spontaneously simulating human conditions. 2. Longer life span of the animal allows more accurate modeling especially in degenerative diseases. 3. Body size and weight more comparable to human 4. Skeletal system more comparable to human 5. Easy management of surgical manipulation 6. More probable clinical translation due to more accurate dosage and drug distribution 7. Ability to use available technology as MRI, CT and physiological monitoring
Disadvantages	1. No anatomical similarity with human body 2. Difficult surgical manipulation 3. Difference in body mass and drug metabolism	1. High cost 2. Higher ethical considerations 3. Longer reproductive cycle 4. Higher cost of maintenance 5. Larger facility requirement 6. Limited ability of genetic manipulation and production of transgenic strains

10.3.3 Challenges in animal experiments

Animal experiments in translational medicine face several notable challenges, namely, animal choice, animal modeling standardization, and interpretation of experimental results. By far, the most difficult challenge is the interpretation of results for application in human trials. Interpretation must incorporate the differences in anatomy, physiology, or metabolism between the examined species and human beings.

10.3.4 Ethical considerations

All animal procedures must follow the ethical guidelines for animal manipulation. These guidelines are summarized in the following main principles:

1. Animal experiments should be conducted only if it the information needed cannot be obtained from other methods such as in vitro biological systems or mathematical modeling.
2. The number of animals used should be reduced to the minimum needed to obtain reliable data.
3. The study procedure must be designed to minimize unfavorable impact on animals. The possible harm to the animals should be carefully balanced with the expected scientific benefit. This rule should be followed in all steps of the experiment: anesthesia, caging, breeding, and transportation in addition to the core steps of the experiment.
4. Animal housing and facilities must provide the best possible animal care.
5. Personnel working with animals should be trained on both the experimental procedure and appropriate and ethical animal handling.

10.4 Clinical trials

After careful consideration of the results of animal studies, clinical trial setting should follow the standard stepwise phases: pilot and phases I, II, III, and IV. Detailed protocol and patient consent are mandatory.

A pilot study is a small-scale study that tests the feasibility of the experimental procedure to be used in a larger-scale study. Results of the pilot study are not representative of the efficacy or safety of the procedure. Pilot studies assess the feasibility of all steps of the procedure: recruitment, labeling, randomization, intervention procedure, and assessment methodology.

Phase I clinical study, previously known as first-in-human study, is a small-scale study conducted on a small number of healthy volunteers to assess the safety of the procedure. Phase I trials are further divided into phase Ia (single dosing) and phase Ib (multiple dosing).

Phase II clinical trials are efficacy studies involving larger number of subjects up to 300 who receive the treatment. It is further divided into phase IIa "proof-of-concept study" and phase IIb "dose-determination study." Study design of phase II trials can be case series or randomized clinical studies.

Phase III studies are designed to test the effectiveness of the procedure in clinical practice. Thus, it compares the new treatment with a standard treatment and involves a large number of centers—"multicenter." It recruits a large number of patients and is typically done over a longer time scale.

Phase IV is a postmarketing study or a confirmatory study involving long-term followup of patients.

Further reading

Corey S, Lippert T, Borlongan CV. Translational lab-to-clinic hurdles in stem cell therapy. Chin Neurosurg Jl 2016;2:41. Available from: https://doi.org/10.1186/s41016-016-0058-z.

Ericsson Aaron C, Crim Marcus J, Franklin Craig L. A brief history of animal modeling. Mo Med 2013;110(3):201–5 May-Jun.

Festing S, Wilkinson R. The ethics of animal research. Talking Point on the use of animals in scientific research. EMBO Rep 2007;8(6):526–30. Available from: https://doi.org/10.1038/sj.embor.7400993.

Hardouin SN, Nagy A. Mouse models for human disease. Clin Genet 2000;57:237–44.

Hickman DL, Johnson J, Vemulapalli TH, Crisler JR, Shepherd R. Commonly used animal models. Princ Anim Res Graduate Undergrad Stud 2017;117–75. Available from: https://doi.org/10.1016/B978-0-12-802151-4.00007-4.

Horvath S, Kemeny A, Pinter E, Gyulai R. A localized aldara (5% imiquimod)–induced psoriasiform dermatitis model in mice using finn chambers. Curr Protoc Pharmacol 2020;90(1):e78.

Leon AC, Davis LL, Kraemer HC. The role and interpretation of pilot studies in clinical research. J Psychiatr Res 2011;45(5):626–9. Available from: https://doi.org/10.1016/j.jpsychires.2010.10.008.

Lilley Elliot, Armstrong Rachel, Clark Nicole, Gray Peter, Hawkins Penny, Mason Karen, et al. Refinement anim model sepsis septic shock. Shock April 2015;43(4):304–16. Available from: https://doi.org/10.1097/SHK.0000000000000318.

Mouse models of human disease. Part II: Recent progress and future directions. Genes Dev 1997;11:11–43.

Naderi MM, Sarvari A, Milanifar A, Boroujeni SB, Akhondi MM. Regulations and ethical considerations in animal experiments: international laws and islamic perspectives. Avicenna J Med Biotechnol 2012;4(3):114–20.

Reza Khorramizadeh M, Saadat F. Animal models for human disease. Anim Biotechnol 2020;153–71. Available from: https://doi.org/10.1016/B978-0-12-811710-1.00008-2.

Rosenthal N, Brown S. The mouse ascending: perspectives for human-disease models. Nat Cell Biol 2007;9:993–9.

Types and phases of clinical trials. American Cancer Society. 18 August 2020. Retrieved 15 September 2020.

Waldner M, Neurath M. Chemically induced mouse models of colitis. Curr. Protoc. Pharmacol 2009;46:5.55.1–5.55.15 © 2009 by John Wiley & Sons, Inc. 6, 1.

Wolfgang MJ, Golos TG. Nonhuman primate transgenesis: progress and prospects. Trends Biotechnol 2002;20:479–84.

CHAPTER 11

Ethical guidelines of stem cell therapy

Contents

11.1 Introduction 121
  11.1.1 Ethical guidelines for the use of embryonic cells 122
  11.1.2 Ethical concerns 122
  11.1.3 Ethical guidelines for preclinical studies 123
  11.1.4 Principles of ethical guidelines of clinical translation 124
Further reading 125

11.1 Introduction

Research on stem cell and related translational regenerative medicine has provoked controversial reactions from regulatory and scientific bodies. Stem cell research serves two crucial roles in scientific research: laboratory research in disease modeling and drug development and a therapeutic role in regenerative medicine. In the first role, only stem cell acquisition is subject to ethical regulation, while in the second, stem cell manipulation is also monitored.

As stem cell therapy provides potential hope for desperate and hopeless patients, it has expanded at a pace that does not match with its scientific research and legislation. The asynchronous growth of the scientific and regulatory parts of stem cell research has resulted in continuous dispute over stem cell applications.

Thus, until now, and more than two decades after the advent of this new field, various governmental and nongovernmental bodies have been striving to align various ethical codes to come up with a universal code for this subject.

Ethical regulation of stem cell research governs both basic research and the translational process. This comprises personnel, venue, process, and tissue. In Chapter 10, we discussed personnel, venue, and process. In this chapter, we will discuss the tissue and the overall translational issues.

Stem Cell Therapy
DOI: https://doi.org/10.1016/B978-0-12-821569-2.00004-1

11.1.1 Ethical guidelines for the use of embryonic cells

Embryonic stem cells represent the core of ethical dilemma in stem cell research. The fact that they are derived from fertilized ova, with potential formation of human beings, exaggerates concerns about misconduct and misuse. The principal guidelines for use of embryonic stem cells in research are:

1. Only surplus embryos from in vitro fertilization procedures are used, with prohibition of fertilization of embryos for the sake of research.
2. Embryos, oocytes, sperms, or other human tissues should not be subjects of commercialization.
3. Embryos donated for research should not be implanted afterward in uterus of humans or animals.
4. Detailed informed consent should be taken from parents without any pressure regarding any fertility services provided.
5. Elimination of any conflict of interest between stem cell researchers and fertility therapists.

11.1.2 Ethical concerns

1. *Source*: embryonic, fetal.

   Authorized stem cells sources are:

   a. Autologous adult stem cells from any tissue: bone marrow, peripheral blood, adipose tissue, or tissue biopsy.
   b. Allogenic adult stem cells from any tissue after proper consent.
   c. Cord blood and tissue after proper consent.
   d. Embryonic stem cells: aborted embryos or surplus embryos from IVF centers. Regulation for using embryonic stem cells from IVF: stem cells are not allowed to be taken from embryos more than fourteen days, embryos should be initially created for fertility not research, embryos are in excess of clinical need or not suitable for implantation and would otherwise be discarded, and lastly, consent of the parents is mandatory.

2. *Gene manipulation*:

   Regulations for gene manipulation involves both induced pluripotent stem cells IPs and somatic cell nuclear transfer (SCNT).

   IPs should not be combined with or added to human or non-human embryos. IPs regulations is variable between countries, with most liberal legislations in Japan.

SCNT stem cells involve transfer of a somatic cell nucleus into enucleated oocyte producing embryonic-like stem cells. Ethical guidelines of SCNT are:

a. Using SCNT stem cells for therapeutic cloning should only be allowed to survive for 14 days.
b. Using SCNT for human productive cloning is prohibited.
c. Formation of SCNT into animal egg is allowed for research purposes, but not longer than 14 days.
d. Formation of SCNT from animal nucleus into human egg is prohibited.

3. *Consent*:

The consent process is an important part of ethical regulations of stem cell research. The consent process should be completely free and informed. Freedom to give consent implies that the participant's decision does not influence his/her right for future therapy. All information that can influence decision-making must be given and explained thoroughly to participants. Confidentiality of all genetic and infectious information of the material should be guaranteed, avoiding obtaining, saving, and using unnecessary information.

4. *Ownership of stem cells / trade prohibition of human tissues*:

Commercialization of human cells, tissues, or products is prohibited. This means that participants should not receive any benefit from donating tissues or cells, whether this benefit is direct financial support or providing therapy at a low cost or similar incentives (Table 11–1).

11.1.3 Ethical guidelines for preclinical studies

1. *Use of experimental animals*:

Experimental animal models should be used in order: small animals models followed by large animal models and then primates. Small animal models are used to explore the stem cell type, clinical efficacy, the mechanism of regeneration, and the route of stem cell injection.

Large animal models should be used in diseases where small animals are not representative of human disease, such as skeletal disorders. Using primates should only be done where other animals are not sufficient and should be supervised by expert zoologists.

2. *Criteria for "proof-of-concept" studies*:

Preclinical research involving animals should be used to answer all questions necessary for clinical translation and to provide adequate

Table 11–1 Ethical issues at different phases of stem cell research.

Laboratory phase	Preclinical phase	Clinical translational phase
1. Source of stem cells a. Embryonic stem cells - Generated during in vitro fertilization procedure not for research - Early stages - Surplus embryos - Consent b. Fetal cells c. Induced pluripotent stem cells d. Cord blood stem cells e. Adult stem cells 2. Avoid cloning of human cells 3. Avoid experiments on germ cells	1. Follow regulations for ethical use of animals in scientific research	1. Using ethically approved stem cell source 2. Follow rules of clinical trials 3. Careful monitoring and followup

safety results to start clinical trials. Thus, the "proof-of-concept" animal experiments should document stem cell type, dose, route, efficacy, mechanism, and safety.

11.1.4 Principles of ethical guidelines of clinical translation

1. *Independent review and oversight*

A close review of all steps of clinical translation is essential. The review should be undertaken by experts in the field and should involve all scientific and technical points in the research. Interpretation of preclinical data to implement in trial design is crucial. There are conditions where the review committee should call a termination or temporary halt of the trial if one of the safety criteria is under concern.

"Clinical hold" is done to protect human participants from unreasonable or significant risk or if the study protocol is deficient in design and fails to meet is objectives.

2. *Voluntary informed consent*

Criteria for free and informed consent were previously outlined for tissue and cell donation. In the clinical trial setting, further explanation of the possible mechanism of action and complications should be emphasized.

3. *Medical monitoring and adverse event reporting*

   Transparent patient monitoring should be done by a third party independent, preferably blinded clinical observer. The occurrence of major adverse events is a cause for trial halt or termination.

4. *Medical innovation*

   In exceptional medical situations, clinicians are allowed to use experimental procedures in a very small number of patients with rigorous oversight. This must be accompanied by a close patient followup, transparency, and an independent peer review. Progression to formal clinical trials should be done as soon as possible.

Further reading

Afshar L, Aghayan HR, Sadighi J, et al. Ethics of research on stem cells and regenerative medicine: ethical guidelines in the Islamic Republic of Iran. Stem Cell Res Ther 2020;11:396. Available from: https://doi.org/10.1186/s13287-020-01916-z.

Caulfield T, Scott C, Hyun I, Lovell-Badge R, Kato K, Zarzeczny A. Stem cell research policy and iPS cells. Nat Methods 2010;7(1):28–33. Available from: https://doi.org/10.1038/nmeth.f.282 Jan.

Farajkhoda T. An overview on ethical considerations in stem cell research in Iran and ethical recommendations: a review. Int J Reprod Biomed 2017;15(2):67–74.

Guidelines for stem cell therapy. Medical development division, Ministry of Health, Malaysia, 2009.

Hyun, et al. ISSCR guidelines. Cell Stem Cell 2008;3:607–9.

Informed Consent Standard for Stem Cell-Based Interventions Offered Outside of Formal Clinical Trials. ISSCR Version 1.0, 12 August 2019.

Kimmelman J, Heslop H, Sugarman J, Studer L, Benvenisty N, Caulfield T, et al. The new ISSCR guidelines: clinical translation of stem cell research. Lancet 2016;387 (10032):1979–81.

CHAPTER 12

Stem cell therapy: practical examples

Contents

12.1 Neuroregeneration 128
  12.1.1 In vitro experiments 128
12.2 Experimental animal study 130
  12.2.1 Induction of animal model 130
  12.2.2 Stem cell therapy 131
  12.2.3 Followup 131
12.3 Results 131
  12.3.1 Clinical translational study 132
12.4 Results of clinical translational study 132
12.5 Replacement of defective protein 134
  12.5.1 Epidermolysis bullosa 134
12.6 Results 135
12.7 Multifactorial defects 135
  12.7.1 Results 136
12.8 Conclusion 137

In this chapter, we will review the regenerative trials in a number of diseases spanning the different defects described in previous chapters: cellular loss, deficient protein, and combined defects. In each example, we will explain the milestones of the research on regenerative therapy from "benchtop to bedside."

Cellular loss

As an example for cellular loss, neuroregeneration following spinal cord injury will be discussed.

Replacement of deficient protein

Epidermolysis bullosa (EB) will be discussed as an example of replacement of deficient protein.

Combined defect

Stem cell therapy for vitiligo will explain the role of stem cells in combined defects.

Multifactorial (Vitiligo)

Stem Cell Therapy
DOI: https://doi.org/10.1016/B978-0-12-821569-2.00011-9

12.1 Neuroregeneration

Stem cell therapy for restoration of neural function offers hope for previously incurable diseases. The idea defies an important "dogma" in classical medical teaching that the neural system "does not regenerate." In addition, the experimental evidence of the presence of more than one pathway in the nervous system inhibiting neural regeneration after injury gave more credibility to this dogma.

Stem cell–induced neural regeneration can present an important therapeutic modality in traumatic neural injury, for example, spinal cord injury and stroke, degenerative neural injury, for example, Alzheimer's disease and Parkinson's disease, as well as genetic neural disorders, for example, spinal muscular atrophy.

In this chapter, we will discuss the steps of implementation of stem cell therapy in the management of traumatic spinal cord injury.

In an attempt to design a stem cell therapy for traumatic spinal cord injury, the study was conducted in three steps: laboratory experiments, preclinical (animal) experiments, and finally, clinical translational experiments.

12.1.1 In vitro experiments

The aims of the in-vitro phase were: Figs. 12–1 and 12–2

1. To verify the ability of mesenchymal stem cells (MSCs) to transdifferentiate into neural cells.
2. To compare the different sources of MSCs for neural transdifferentiation.
3. To establish a detailed laboratory protocol for MSC neural transdifferentiation.

1. Verification of the ability of MSCs to transdifferentiate into neural cells:

a. *Methods*:
   1. Isolation of MSCs from bone marrow, cord blood, Wharton jelly, and adipose tissue was done according to standard protocols (see Appendix). Isolation of hematopoietic stem cells (HSCs) was done from bone marrow, leukapheresis product, and cord blood according to standard protocols (see Appendix).
   2. Isolated stem cell populations were verified as follows: for MSCs, according to the ISCT minimal criteria for MSC identification, and for HSC, using flowcytometric detection of CD34 expression.
   3. Isolated stem cells were subjected to induced neural differentiation using three different laboratory methods:
      a. Coculture with isolated neurospheres from murine brain

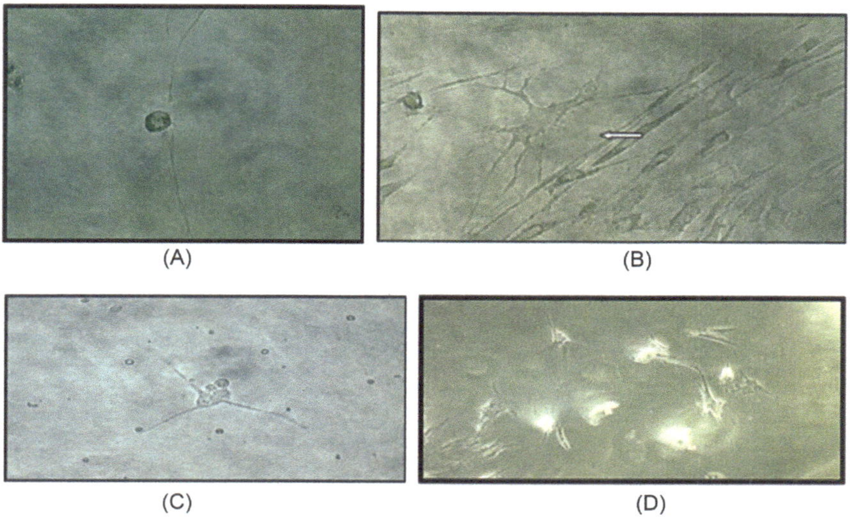

Figure 12–1 (A) Neural differentiation of mesenchymal stem cell showing neural synapse; (B) Neural differentiation of mesenchymal stem cell showing oligodendrocyte; (C) Neural differentiation of mesenchymal stem cell showing morphology of multipolar neuron; (D) Neural differentiation of mesenchymal stem cells showing astrocytes with intercellular connections.

b. Addition of culture supernatant of neurospheres
c. Addition of cytokine cocktail (basic fibroblast growth factor, ascorbic acid, and nerve growth factor).
d. Evaluation of the efficiency of neural differentiation in different conditions. This was done using: morphology and immuncytochemistry as shown in e and f.
e. Morphology: MSC morphology changes gradually to neural morphology: astrocytes, oligodendrocytes, and neuroglial cells.
f. Immunocytochemistry: expression of neural lineage–specific markers, for example, nestin expression, β-tubulin III expression, glial fibrillary protein GFAP, and neurofibrillary protein. In this experiment, we used nestin and GFAP expression.

2. Comparison between different lineages (HSC/MSC) and different sources (BM/CB, AD) of stem cells.
3. Comparison between different laboratory conditions and establishment of the best laboratory protocol for induction of neural differentiation.

Data verified from laboratory experiment and used to proceed to preclinical stage:

The results of comparative experiments proved superior neural trans-differentiation of MSCs than HSCs. Evaluation of the effect of the source

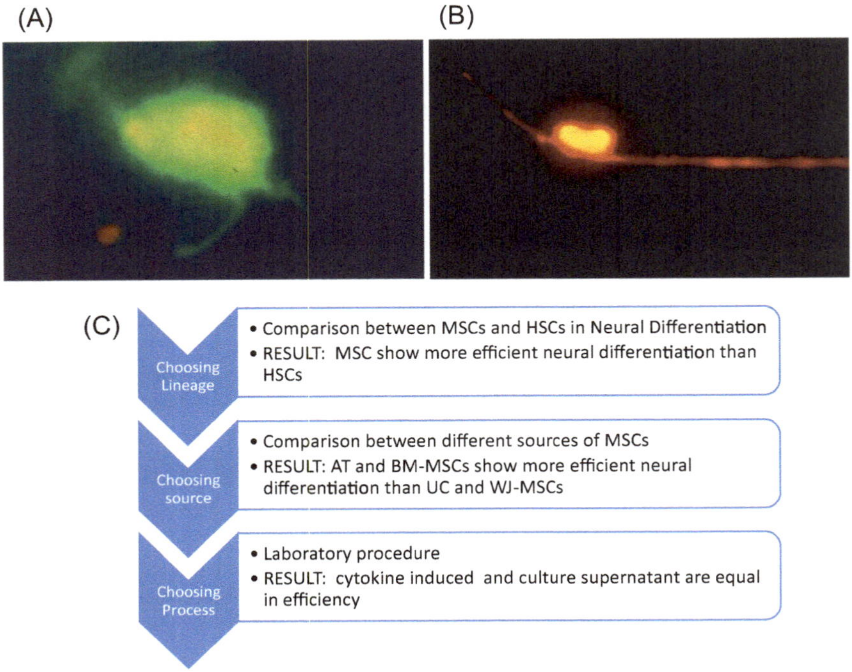

Figure 12–2 (A) Nestin-positive neurospheres; B. GFAP-positive neural-differentiated mesenchymal stem cells; C. Procedure of stem cell therapy.

of MSCs on neural transdifferentiation showed that MSCs isolated from different sources show variable neural differentiation capacity, with adipose- and bone marrow–derived MSCs showing higher transdifferentiation percentage as well as more functional neurons than umbilical cord or Wharton jelly–derived MSCs.

Experiments testing the effect of various of induction of transdifferentiation showed comparable results with cytokines and culture supernatant.

12.2 Experimental animal study

12.2.1 Induction of animal model

Large animal (canine) model was chosen as it is preferred in simulation of locomotor and skeletal defects. Induction of the spinal cord injury was done using surgical crushing of the cord at the level of L4, leading to paraplegia.

Clinical evaluation of the effect of the procedure was done. Animals experienced loss of sensation and motor power below the level with incontinence to urine and stools.

A sample of the animal groups was sacrificed to evaluate the histological changes following the injury.

12.2.2 Stem cell therapy

The canine model group was subdivided into subgroups in order to study the effect of the timing of stem cell injection, dose, and the route of injection on the regenerative effect of stem cells. A group was added that received cell-free exosomes instead of stem cells. Injected cells were labeled with green fluorescent protein in order to track the cells in histological sections.

12.2.3 Followup

1. Clinical scales: all study groups were followed up by clinical examination and Olby score determination for six months following the injection.
2. Histological study: at the end of the followup period, animals were sacrificed to compare the histological changes occurring in different subgroups with those of sham-injected controls, as well as normal animals.
3. Cell tracking: cells were tracked to study their distribution as well as in vivo differentiation.

12.3 Results

Data verified from animal study and used to proceed to clinical translation: Fig. 12–3

1. Injected stem cells were safe with no immediate or short-term complications.
2. Injected undifferentiated stem cells were homed and were incorporated into neural tissue as evidenced by green fluorescent protein staining.
3. Stem cells induced clinically detectable improvement in animal models.
4. Undifferentiated stem cells showed the same clinical improvement as the differentiated stem cells.
5. Injected stem cells induced clinically and histologically evident neuroregeneration.
6. Dose of stem cells used was sufficient; however, dose escalation studies may be done.
7. Cell-free exosomes induced neural regeneration clinically comparable to stem cell injection.

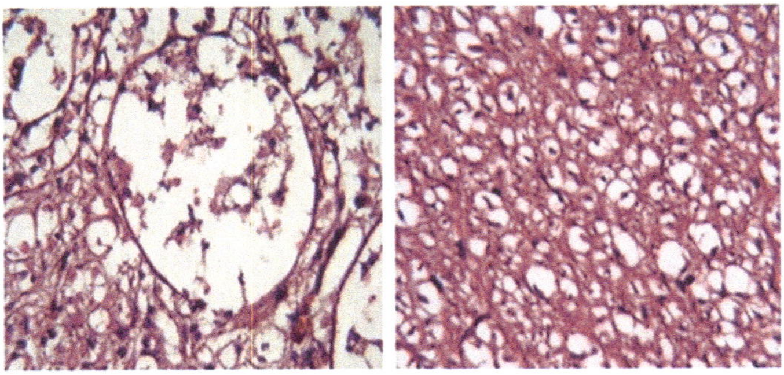

Figure 12–3 Histological evaluation of spinal cord sections showing maceration and vacuolation in the control group and histological regeneration in the stem cell injection group.

12.3.1 Clinical translational study

Clinical translation was started after completing animal study. Clinical trial was registered:

ClinicalTrials.gov Identifier: NCT00816803

Clinical trial proceeded in the following order:

1. Pilot trial for safety: recruited 30 patients randomly and followed up for immediate and short-term complications.
2. Phase I/II: recruited 100 patients with cervical and thoracic spinal cord injuries.

Patients were selected according to the following criteria: inclusion criteria are chronic traumatic spinal cord injury of 10–36 months duration with no neurological recovery for the last six months and the absence of concomitant systemic disease.

Patients with nontraumatic or acute spinal cord injury or with major impairment of respiratory or cardiac functions were excluded.

Baseline assessment was performed using the American Spinal Cord Association (ASIA) Impairment Scales A and B. Patients received intrathecal injection of autologous MSCs as well as physiotherapy and were followed for 12 months. Followup was done using clinical examination, somatosensory evoked potential, and MRI imaging Figs. 12–4 to 12–6.

12.4 Results of clinical translational study

Phase I: Phase I study, following patients for six months, proved the safety and tolerability of the procedure.

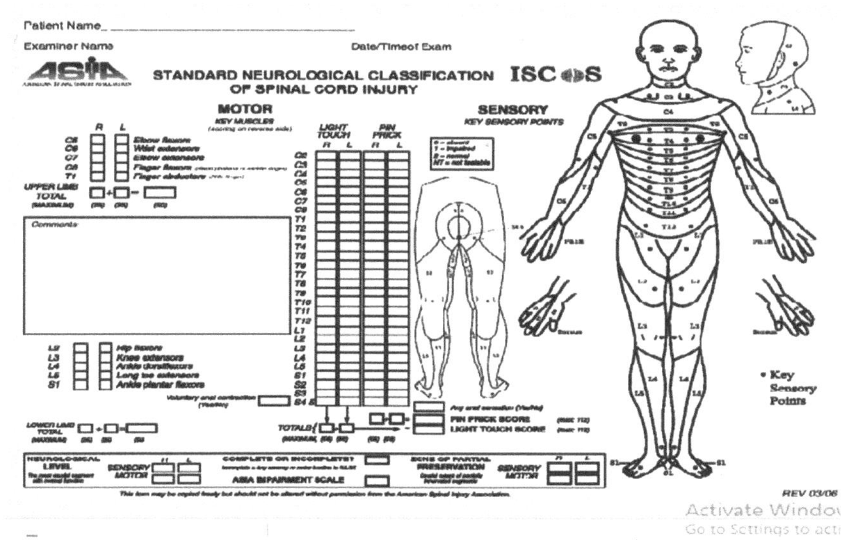

Figure 12–4 American Spinal Cord Association score.

ASIA Impairment Scale	*Description*
A	Complete. No sensory or motor function is preserved in the sacral segments S4–S5.
B	Incomplete. Sensory but not motor function is preserved below the neurological level and includes the sacral segments S4–S5.
C	Incomplete. Motor function is preserved below the neurological level, and more than half of key muscles below the neurological level have a muscle grade less than 3 (grades 0–2).
D	Incomplete. Motor function is preserved below the neurological level, and at least half of key muscles below the neurological level have a muscle grade greater than or equal to 3.
E	Normal. Sensory and motor function are normal.

Figure 12–5 American Spinal Cord Association scale.

Phase II/III: Cumulative data from multicenter study showed collective improvement in 43.2% of cases, 40.2% improvement in sensory improvement, 16% improvement in motor scores, and 8% improvement in sphincteric control. The improvement was more prominent in thoracic injuries. The resultant change in quality of life was evidenced by the change in ASIA grade and functional ambulatory categories. Sensory improvement

Score	Category	Interpretation
0	Nonfunctional ambulator	
1	Ambulator, dependent on physical assistance - level I	Indicates a patient who requires continuous manual contact to support body weight as well as to maintain balance or to assist coordination.
2	Ambulator, dependent on physical assistance - level II	Indicates a patient who requires intermittent or continuous light touch to assist balance or coordination.
3	Ambulator, dependent on supervision	Indicates a patient who can ambulate on level surface without manual contact of another person but requires standby guarding of one person either for safety or verbal cueing.
4	Ambulator, independent level surface only	Indicates a patient who can ambulate independently on level surface but requires supervision to negotiate (e.g. stairs, inclines, nonlevel surfaces).
5	Ambulator, independent	Indicates a patient who can walk everywhere independently, including stairs.

Figure 12–6 Functional ambulatory categories.

was evidenced by improvement in 1–6 dermatomes with a mean of 2.85 dermatomes. Sensory improvement was in both pin prick and light touch.

Sphincteric improvement was in the form of change from urinary incontinence to reflex voiding with or without sensation. None of the treated groups reached complete continence. Regarding bowel continence, most of the improved cases developed partial control with or without sensation.

12.5 Replacement of defective protein

12.5.1 Epidermolysis bullosa

EB, or butterfly disease, is a rare genetic disorder that affects over half a million patients worldwide. The disease is characterized by easy blistering of the skin and mucous membranes due to mutations leading to defective synthesis in anchoring proteins between the layers of the skin. Clinical spectrum of EB varies from mild cases, which usually symptomatize as the patient starts to crawl or walk, to severe complicated cases, which may suffer from esophageal narrowing, severe hand and foot deformities, and squamous cell skin carcinoma.

There is no cure for EB, and management is usually symptomatic wound care and treatment of complications.

Research strategy for stem cell treatment of EB was as follows:

Twelve EB patients, diagnosed clinically and by skin biopsy, were enrolled.

1. Initial evaluation of patients: evaluation was done to provide a quantitative data of the number of blisters and duration in addition to the general condition of the patient (weight, height, and growth percentile)

2. Initial evaluation of donors: potential donors were evaluated using history and skin biopsy. All patients received stem cells from one of their parents.
3. Stem cell preparation: Bone marrow was aspirated, and MSCs were isolated from the sample.
4. Stem cell infusion: MSCs were infused intravenously in a dose of 2 million/kg.
5. Followup: it was done by comparing the frequency and number of blisters, rate of healing of new blisters, as well as electron microscopic examination of skin biopsy.

12.6 Results

Haploidentical MSC infusion was well tolerated with no reported side effects or complications. Clinical evaluation of the patients showed a significant decrease in the rate of appearance of new blisters with faster healing of erupted ones. Electron microscopic examination of skin biopsies revealed increased number of anchoring fibrils after treatment. Fig. 12–7

12.7 Multifactorial defects

Vitiligo is a multifactorial disease involving defective melanin pigmentation with autoimmune features. The principle of regenerative therapy for vitiligo involves:

1. Replacement of defective melanocytes:

   Replacement of defective melanocytes can either be done through isolation of melanocyte stem cells from skin biopsy and further propagation of this population; or transdifferentiation of bone marrow–derived HSCs.

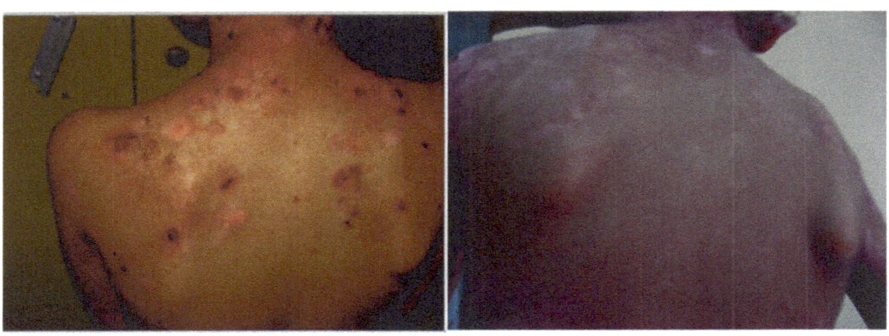

Figure 12–7 Improvement in skin eruptions after stem cell therapy.

This approach requires local intralesional injection of melanocytes.

2. Immune regulation:

   Systemic injection of autologous MSCs.

12.7.1 Results

12.7.1.1 In vitro experiments

Isolation of melanocyte stem cells from skin biopsy was possible. However, further in vitro expansion was very slow and difficult. Thus, reaching acceptable number of melanocytes from skin biopsy was impractical.

Transdifferentiation of bone marrow–derived HSCs to melanocytes was possible. Expansion of melanocytes was difficult; however, reaching acceptable number of melanocytes was possible using a large initial bone marrow sample.

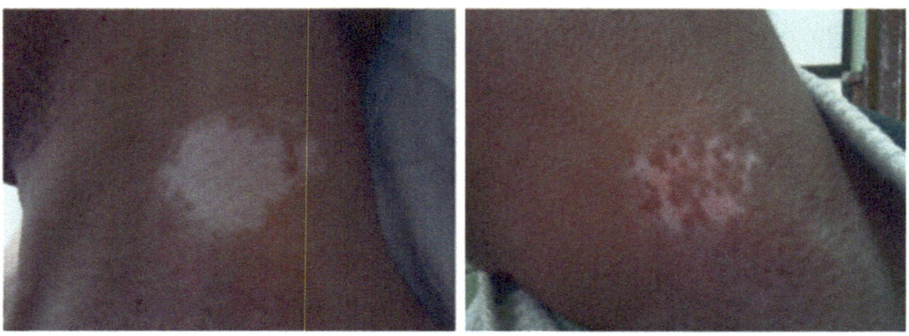

Figure 12–8 Repigmentation following local injection of transdifferentiated melanocytes.

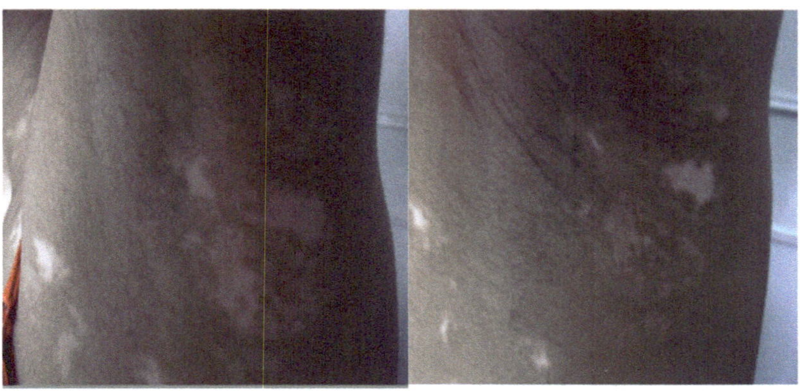

Figure 12–9 Partial regimentation following systemic mesenchymal stem cell.

12.7.1.2 Clinical results

Repigmentation of vitiliginous patches occurred in locally injected lesions and partially in noninjected lesions due to improvement in the autoimmune dysfunction. Figs. 12–8 and 12–9

12.8 Conclusion

Stem cell therapy is a promising solution for correcting degenerative, traumatic, genetic, and autoimmune disorders. A systematic approach is necessary to design a therapeutic strategy. Preliminary positive results are encouraging, and further clinical applications will shape the future of medicine.

Further reading

Abdelaziz Osama S, Marie Ahmed, Abbas Mohamed, Ibrahim Mohamed, Gabr Hala. Feasibility, safety, and efficacy of directly transplanting autologous adult bone marrow stem cells in patients with chronic traumatic dorsal cord injury. A pilot clinical study. Neurosurg Q 2010;20:216–26.

Abo El-Kheir W, Gabr H, Awad MR, von Wild K, Ramadan M. Autologous mesenchymal stem cells for neuro-regeneration after traumatic spinal cord injury: a comparison between bone marrow and peripheral blood populations. Am J Neuroprot Neuroregen 2010;2:1–8.

Ahmedy E, Kandel S, Gabr H, Rizk S, Khalifa K, Kamal S. Neurogenic differentiation of bone marrow-derived mesenchymal stem cells using neural induction medium: a morphological and histochemical study. Am J Biosci Bioeng 2015;3(4-1):43–50. Available from: https://doi.org/10.11648/j.bio.s.2015030401.17.

Ahmedy E, Kandel S, Gabr H, Rizk S, Khalifa K, Kamal S. Neurogenic differentiation of bone marrow-derived mesenchymal stem cells using neural induction medium: a morphological and histochemical study. Am J Biosci Bioeng 2015;3(4-1):43–50.

Aly A, Mohsen l, Badrawi N, Gabr H, Ali Z, Akmal D. Viability and neural differentiation of mesenchymal stem cells derived from umbilical cord following perinatal hypoxia. J Perinatol 2011;1–6.

Aly H, Mohsen L, Badrawi N, Gabr H, Ali Z, Akmal D. Viability and neural differentiation of mesenchymal stem cells derived from the umbilical cord following perinatal asphyxia. J Perinatol 2012;32:671–6. Available from: https://doi.org/10.1038/jp.20110.174.

Ayad H, Zahra M, Gabr H, Ismaeil A, Abo El-Kheir Y. The role of adipose derived mesenchymal stem cells in enhancement of peripheral nerve injuries repair. Am J Biosci Bioeng 2015;3(4-1):18–23.

El-Darouti M, Fawzi M, Amin I, Abdel-Hay R, Hegazy R, Gabr H, et al. Treatment of dystrophic epidermolysis bullosa with bone marrow non-hematopoeitic stem cells: a randomized controlled trial. Dermatol Ther 2016;29(2):9–100 March-April.

El-Darouti M, Fawzy M, Amin I, Abdel Hay R, Hegazy R, Gabr H, et al. Treatment of dystrophic epidermolysis bullosa with bone marrow non-hematopoeitic stem cells: a randomized controlled trial. Dermatol Ther 2016;29(2):96–100. Available from: https://doi.org/10.1111/dth.12305 Epub 2015 Oct 5. PMID: 26439431.

El-Kheir WA, Gabr H, Awad MR, Ghannam O, Barakat Y, Farghali HA, et al. Autologous bone marrow-derived cell therapy combined with physical therapy induces functional improvement in chronic spinal cord injury patients. Cell Transpl 2013; Feb 26.

Gabr H, Ghannam O, Awad R, Von Wilde K, Abo El-Kheir W, Ewes I. Autologous mesenchymal stem cell therapy for spinal cord injury: long term safety and clinical efficacy. Am J Neuroprotec Neuroregen 2011;3(1):100–6.

Gabr H, El-Kheir WA, Farghali HA, Ismail ZM, Zickri MB, Maadawi ZM, et al. Intrathecal transplantation of autologous adherent bone marrow cells induces functional neurological recovery in a canine model of spinal cord injury. Cell Transpl 2014; Jul 15.

Gabr H, El-Kheir WA, Ghannam O, El-Fiki ME, Salah Y. Intrathecal autologous bone marrow derived MSC therapy in cerebral palsy: safety and short term efficacy. Am J Biosci Bioeng 2015;3(4-1):1–6.

Gabr H, El-Kheir WA, Ghannam O, El-Fiki ME, Salah Y. Intrathecal autologous bone marrow derived MSC therapy in cerebral palsy: safety and short term efficacy. Am J Biosci Bioeng 2015;3(4-1):24–9.

Ismail Zeinab M, Abdel-Dayem Menna M, Salama Nagla M, Gabr Hala, Ismail Dalia I, Abdelhafiz Ahmed S, et al. Two-step induction of dopaminergic neuron differentiation of leukapharesis-derived mesenchymal stem cells. Am J Biosci Bioeng 2015;3(4-1):7–17.

Kishk NA, Gabr H, Hamdy S, Afifi L, Abokresha N, Mahmoud H, et al. Case control series of intrathecal autologous bone marrow mesenchymal stem cell therapy for chronic spinal cord injury. Neurorehabil Neural Repair 2010;24(8):702–8. Available from: https://doi.org/10.1177/1545968310369801 Epub 2010 Jul 26. PMID: 20660620.

Kishk Nirmeen A, Gabr Hala, Hamdy Sherif, Afifi Lamia, Abokresha Noha, Mahmoud Hayam, et al. Case control series of intrathecal autologous bone marrow mesenchymal stem cell therapy for chronic spinal cord injury. Neurorehabil Neural Repair Online First 2010.

Nirmeen A, Kishk a, Noha T, Abokrysha a, Gabr Hala. Possible induction of acute disseminated encephalomyelitis (ADEM)-like demyelinating illness by intrathecal mesenchymal stem cell injection. J Clin Neurosci 2013;20:310–12.

Radhakrishnan S, Trentz O, Reddy M, Rela M, Kandasamy M, Sellathamby S. In vitro transdifferentiation of human adipose tissue-derived stem cells to neural lineage cells - a stage-specific incidence. Adipocyte 2019;8(1):164–77.

Sheng SR, Wang XY, Xu HZ, Zhu GQ, Zhou YF. Anatomy of large animal spines and its comparison to the human spine: a systematic review. Eur Spine J Off Publ Eur Spine Soc Eur Spinal Deformity Soc Eur Sect Cerv Spine Res Soc 2010;19(1):46–56. Available from: https://doi.org/10.1007/s00586-009-1192-5.

Urrutia DN, Caviedes P, Mardones R, Minguell JJ, Vega-Letter AM, Jofre CM. Comparative study of the neural differentiation capacity of mesenchymal stromal cells from different tissue sources: an approach for their use in neural regeneration therapies. PLoS One 2019;14(3):e0213032. Available from: https://doi.org/10.1371/journal.pone.0213032.

Zickri MB, El Aziz DH, Metwally HG. Histological experimental study on the effect of stem cell therapy on adriamycin induced chemobrain. Metwally Int J Stem Cell 2013;6:2.

Appendix A

A.1 Standard operating procedures for stem cell therapy procedures

A. **Collection:**
   1. Peripheral blood hemopoietic stem cells (HSCs)
   2. Bone marrow HSCs
   3. Cord blood HSCs
   4. Adipose tissue

B. **Isolation:**
   1. HSCs
   2. Mesenchymal stem cells (MSCs)

C. **Characterization**

D. **Expansion**

A.1.1 Collection

A.1.1.1 Collection of peripheral blood hemopoietic stem cells

HSCs are collected from peripheral blood using mobilization followed by leukapheresis.

Leukapheresis, a specific type of apheresis, is a laboratory procedure for separating a specific population of white cells from the whole blood of a patient. In the stem cell setting, mononuclear cell population is separated as it is the morphological entity of stem cells. HSCs are mobilized from the bone marrow to harvest an appropriate dose of stem cells.

1. Stem cell mobilization:

   During steady state, circulating HSCs are less than 0.001% of the white cell population. Therefore, induction of stem cell mobilization is mandatory before leukapheresis. Standard mobilization is one using G-CSF in a dose of 5 mcg/kg/day in subcutaneous dose daily for 5 days. There is great variation in the response to G-CSF. The main determining factor is white cell count of at least 30×10^9/L. Failure to reach this count will necessitate either the addition of chemotherapy or CXCR4/CXCL12 pathway blocker, plerixafor, or stem cell factor.

2. Intravenous lines: either two lines, one for removal of blood and the other returning other components, or a central venous line is inserted.

3. Leukapheresis takes 2–3 hours in average to collect a stem cell dose of $2–3 \times 10^6$/kg.

4. Possible complications: Leukapheresis is a well-tolerated procedure. However, some people experience cold sensation, numbness, and painful muscle spasms. This is usually due to hypocalcemia and treated by calcium administration.

A.1.1.2 Collection of bone marrow hemopoietic stem cells

Even though peripheral blood stem cell collection is taking more ground than bone marrow, some cases require bone marrow-stem cell therapy.

Under general anesthesia, in an operating room, bone marrow is harvested from the posterior iliac spine. Harvesting is performed from multiple sites, 5 mL from each puncture site. The syringes used should be heparinized.

The volume required to achieve optimum CD34 + dose is about 10–15 mL/kg body weight, that is, 800–1000 mL.

The marrow collected is injected into special bone marrow collection bag of volume 1500 mL, with acid citrate dextrose anticoagulant. The collection bag contains prefilter to prevent particles from flowing to the patient.

A.1.1.3 Cord blood stem cell collection

Collection of stem cells from cord blood can be done either in utero or ex utero. In utero collection is done after delivery of the baby and before delivery of the placenta. Ex utero collection is done after delivery of the placenta.

Collection is either done into a special cord blood collection bag or using heparinized large syringes.

In both cases, the umbilical artery is cannulated to start collection.

Average blood collected ranges from 50 to 70 mL.

A.1.1.4 Collection of stem cells from adipose tissue

Adipose-derived stem cells can be collected from fat tissue through direct excision or lipoaspirate. Lipoaspiration is a procedure that removes fat tissue surgically through suction of subcutaneous fat. Adipose-derived stem cells are a constituent of the stromal vascular fraction. The lipoaspirate is incubated for 30–60 minutes at 37°C with collagenase, centrifuged for 20 minutes at 1800 rpm, and the stromal vascular fraction is pelleted at the bottom of the tube.

A.1.2 Isolation

A.1.2.1 Mononuclear cell separation by density gradient centrifugation

Background:

Density gradient separation techniques are the first step of most cell separation procedures. The standard (1077 g/L) is used for lymphocyte separation since stem cells are morphologically similar to lymphocytes. Other cell populations, for example, monocytes, eosinophils, and neutrophils, require different density gradients. Density gradient separation may not yield as many cells as the simple lysis methods. Simple lysis yields collective white cell population.

A. Equipment:

1. Centrifuge with swing-out rotor and brake switched off
2. Sterile pipettes
3. Sterile falcon tubes

B. Reagents:

1. Ficoll density gradient solution
2. Phosphate buffered saline

C. Preparation of sample:

1. Collect blood sample under aseptic conditions on preservative-free heparin.
2. Add phosphate buffered saline (PBS) in a ratio of 1:1 and mix by inverting the tube several times.
3. Bring Ficoll-Paque solution to room temperature; mix by inverting the bottle several times.
4. Add 3 mL Ficoll-Paque solution to a sterile falcon tube.
5. Carefully layer the diluted blood sample (4 mL) on the solution without mixing.
6. Centrifuge at 400 g for 20–30 minutes at 18°C–20°C with brake turned off.
7. The blood sample will be separated into four layers. Remove the uppermost layer (plasma).
8. To collect the mononuclear cell layer, enter the pipette closed, near the side of the tube and open carefully and collect the layer in a circular manner.
9. The mononuclear cells collected are washed and counted.

A.1.2.2 Mesenchymal Stem Cell Isolation from Wharton Jelly

Methodology:

A. *Explant Culture:*

1. Human umbilical cord is collected immediately after birth, after informed consent, and transported to the laboratory in a sterile container in sterile saline.
2. Under laminar airflow cabinet, the cord is rinsed many times in sterile saline supplemented with 10% antibiotics and 10% antifungal (100 U/mL of penicillin, 100 μg/mL of streptomycin, and 250 ng/mL of amphotericin B).

3. The cord is cut to 2–3 cm pieces and incubated in sterile saline supplemented with antibiotics and antifungal for one hour at 4°C.
4. After 1 hour, the cord in rinsed with PBS, and disinfected by rinsing with 70% ethanol, followed by a final wash with PBS.
5. The cord pieces are placed in sterile petri dish and cut longitudinally to open the cord and expose the Wharton jelly.
6. The three vessels in the cord are dissected and removed using forceps.
7. Scraping of the Wharton jelly is done carefully after removing the three vessels in the cord.
8. The scraped Wharton jelly is rinsed thoroughly by PBS.
9. Two to three pieces of scraped Wharton jelly are placed in T25 tissue culture flasks and left to adhere for 1–2 hours at room temperature.
10. Add warm complete medium (DMEM + 20% fetal calf serum + 20% L-glutamine + 10% antibiotics/antifungal) dropwise carefully to avoid disturbing the adherent tissue.
11. The tissue culture flasks are incubated at 37°C, 5% CO_2 for 48 hours.
12. After 48 hours, the medium is replaced completely, with further half change every 48 hours.
13. The cultured flasks are evaluated for cell expansion from explant, and after 7–10 days, the tissue pieces are removed.
14. When the cells reach 85% confluence, the cells are harvested and passaged.

B. *Tissue digestion*

1. The steps from 1 to 8 are done as above.
2. The scraped Wharton jelly in incubated in trypsin–EDTA for 30 minutes at 37°C.
3. After incubation, 1 mL fetal bovine serum is added to stop the action of trypsin.
4. The sample is filtered using sterile 70 μm filters to remove undigested tissue.
5. The cells are cultured in 5 mL complete medium in T25 flasks.

A.1.3 Mesenchymal stem cell culture

1. Adjust the cell count to 5×10^5/mL
2. Using T25 tissue culture flask, the culture suspension is done as follows:
   a. Cell suspension (5×10^5/mL)
   b. Penicillin/Streptomycin (50 μL, 1%)

c. Antimycotic (50 μL, 1%)
d. FCS (500 μL, 10%)
e. DMEM medium to total volume of 5 mL

3. Tissue culture flasks are incubated at 37°C in a humidified CO_2 incubator (5% CO_2).
4. After 48 hours, the nonadherent cells are discarded, and the medium is changed.
5. Afterward, 2 mL of the medium is changed every three days.
6. Morphological evaluation of the cells is done until cells reach 80%–85% confluence.
7. Harvest:
   a. Adherent cells are gently washed twice with prewarmed PBS.
   b. 1–2 mL of 0.25% trypsin-EDTA (prewarmed) is layered over the cell layer in the flask
   c. The flasks are left at 37°C for 5 minutes.
   d. 2–3 mL of the complete medium is added to stop the action of trypsin.
   e. The cells are washed with PBS and collected in sterile centrifuge tubes.
   f. Cells are centrifuged at 1500 rpm for 10 minutes.
   g. The cell pellet is resuspended in the complete medium. Cell count and viability are performed, and the cells are set in a new culture for passage 1.

A.1.3.1 Identification of mesenchymal stem cells

According to the ISCT guidelines, MSCs are identified by:

1. Plastic adherence
2. Fibroblast morphology
3. Variable expression (≥20%) of at least two of the MSC markers: CD105, CD90, CD73, CD271, and CD44, with negative expression of the myeloid markers CD34 and CD45.
4. Ability for trilineage differentiation: osteogenic, chondrogenic, and adipogenic lineages.

A.1.4 Differentiation

A.1.4.1 Plating cells for differentiation

1. Cells harvested from passages 2–3 are plated with 24-well plates in a concentration of 30–50,000/well in the complete medium.
2. After 24 hours, differentiation medium is added as follows:
   a. Osteogenic medium: 10 mM glycerol phosphate, 100 μg/mL ascorbic acid, and 100 nM dexamethasone.

b. Chondrogenic medium: high glucose DMEM, 100 μM of ascorbate-2-phosphate, and 1 μg.mL of TGFβ1.
c. Adipogenic medium: 10 μg/mL insulin, 100 nM dexamethasone, and 200 μM indomethacin.

3. The differentiation medium is changed every 48 hours.

A.1.4.2 Evaluation of differentiation

Cells are evaluated after 10 days as follows:

a. Biochemical estimation of alkaline phosphatase in osteoblast differentiation.
b. Special stains: the medium is aspirated, the cells are washed using PBS, and fixed for 1 minute. The cells are stained using Alizarin red for osteogenic differentiation, oil red O for adipogenic differentiation, and toluidine blue for chondrogenic differentiation.

A.1.5 Coating of flasks

Coating of flasks is used to either improve cell attachment or selectively enhance a specific cell population. The most common coatings used for improvement of attachment are collagen and gelatin coating. Other specific coating solutions are laminin, fibronectin, and poly-lysine.

Collagen Coating:

1. Collagen is diluted to 50 μg/mL using 0.01 M HCl.
2. Collagen is added to the plate or flask in a concentration of 5–10 μg/cm surface.
3. The plates are incubated for 1 hour at room temperature.
4. The plates are rinsed well with PBS to remove excess collagen and acid.

This can be also used for coverslips to be used inside the tissue culture plates.

Gelatin coating:

1. Gelatin working solution is prepared as 2% w/v solution in tissue culture grade water.
2. The solution is sterilized by autoclaving at 120°C for 30 minutes.
3. The culture surface is covered with 5010 μL gelatin solution/cm^2.
4. The plates are left to dry for 2 hours.

Laminin coating steps:

1. Laminin is thawed at 2°C–8°C
2. Thawed laminin is diluted in balanced salt solution to a concentration of 0.01%.

3. The culture surface is covered completely with laminin.
4. The excess solution is removed immediately.
5. The plates are left to dry for 45 minutes at room temperature before use.

A.1.6 Labeling

Iron Oxide Nanoparticle Labeling:

1. Cells are plated in 24-well plates for 24 hours.
2. Cells are cultured in high glucose DMEM, complete medium as described before.
3. Working solution of ferumoxide is prepared as 11.2 mg Fe/mL solution (particle size 80–150 nm).
4. The solution is added in a concentration of 10 μL/mL and left for 5 minutes.
5. Cells are washed, harvested, and stained for iron using Prussian blue stain.

List of Basic Supplies Needed in Stem Cell Lab:

I. Plasticware:
  1. Sterile disposable Pasteur pipette
  2. Sterile disposable pipettes (1, 2, 5 mL)
  3. Sterile centrifuge tube (15, 50 mL)
  4. Sterile tissue culture flasks (T25, T75)
  5. Sterile tissue culture plates (single, 6-, 12-, 24-, 96 well)

II. Instruments:
  1. Sterile disposable blades
  2. Scissors
  3. Scrapers

III. Media:
  1. Basic media: RPMI, DMEM
  2. High glucose media
  3. L-Glutamine
  4. Trypsin–EDTA
  5. Collagenase
  6. Ficoll separating solution
  7. Antibiotics (penicillin, streptomycin)
  8. Antifungal
  9. Coating solutions: collagen, gelatin, fibronectin, laminin

IV. Growth factors:
  1. Fibroblast growth factor
  2. Transforming growth factor beta
  3. Hepatocyte growth factor
  4. Epidermal growth factor
  5. Insulin-derived growth factor
  6. Activin

V. Chemicals and bioactive peptides:
  1. Ascorbic acid
  2. Dexamethasone

VI. Histochemical Stains:
  1. Trypan blue
  2. Giemsa stain
  3. Alizarin red
  4. Oil red
  5. Prussian blue stain

Index

Note: Page numbers followed by "*f*" and "*t*" refer to figures and tables, respectively.

A

Adipocytes, 10, 13
Adipose stem cells (ASCs), 61–62, 140
Adipose tissue, mesenchymal stem niche in, 15–17
Administrative team, 94
Adventitial reticular cells, 13
Alzheimer's disease, 114*t*, 128
Anatomy and histology
  blood supply, 8–10
  components of hemopoietic stem cell niche, 12–14
  differences between osteoblastic and vascular niches, 12
  hemopoietic stem cell niches, 11–12
  mesenchymal stem cell niche, 14–15
    in adipose tissue, 15–17
    bone marrow, 14–15
  niche, 10–11
Angiogenesis, 44–46
Animal models, 113
  experimental, 123
  methods of inducing, 114–116, 130–131
    biological, 114
    chemical, 114
    gene manipulation, 114–116
    of human diseases, 114*t*
    surgical, 114
Artificially derived pluripotent stem cells, 23
ASCs. *See* Adipose stem cells (ASCs)
Antimicrobial functions, 51
Antiinflammatory effects, 51

B

B cells, 24, 52
Bipotent stem cells, 24
Blood supply, 8–10
BM-HSCs. *See* Bone marrow hemopoietic cells (BM-HSCs)
BM-MSCs. *See* Bone marrow mesenchymal stem cells (BM-MSCs)
BMT. *See* Bone marrow transplantation (BMT)
Bone marrow, 7–8, 8*f*, 9*f*, 58, 63
  biopsy, 9*f*
  innervation, 9–10
  sinusoids, 8–9
  transplantation, 4
  vasculature, 10*t*
Bone marrow hemopoietic cells (BM-HSCs), 62, 140
Bone marrow mesenchymal stem cells (BM-MSCs), 58–61
Bone marrow transplantation (BMT), 4, 74–75
  autopsy of patients undergoing gender-mismatched, 4
  nonhemopoietic effects of, 4
Bone morphogenetic protein (BMP) pathway, 42–43

C

Cadherins, 11*t*, 14
Cell division, 14
Cell dose, 112
Cell function after injection, 112
Cell fusion, 28, 48–49
Cell lineage, 71–73
Cell tracking, 103, 112
Cell type, 74
  hemopoietic stem cells, 74, 74*t*
  mesenchymal stem cells, 74, 75*t*
Cellular defects, 70–71
Cellular loss, 128–130
CFU-S. *See* Colony-forming unit-spleen (CFU-S)
Chemokines, 47
Circadian rhythms, 18
Circulating stem cells, 29, 32

Clean room facilities, 89, 91*t*, 92*f*
Clinical evaluation team, 95
Clinical followup team, 95
Clinical translational study, 132–134
Clinical trials, 117–118
Collagen coating, 144
Collection team, 95
Colony-forming unit-spleen (CFU-S), 3–4
Complement proteins, 31
Consent process, 123
Cord blood, 63–64
 stem cell collection, 140
Cord tissue, 58–61
Cytokines, 18–19, 47, 77–78

D

Degenerative diseases, defects in, 72*t*
Density gradient centrifugation, 140–141
Dermal papilla (DP), 16
Dextran sodium sulphate, 114
Differentiation of stem sells, 29
 evaluation of, 144
 plating cells for, 143–144
DNA methylation, 42

E

EB. *See* Epidermolysis bullosa (EB)
ECM. *See* Extracellular matrix (ECM)
ECs. *See* Endothelial cells (ECs)
Embryonic stem cells, 23, 122
Endosteal lining, 8
Endothelial cells (ECs), 8–9, 13, 45, 63–64
Engraftment, 29–32
Epidermal stem cell niche, 16
Epidermolysis bullosa (EB), 71, 127, 134–135
Epigenetic modification, 41–42
Equipment requirements, 90–94, 93*t*
Ethical guidelines
 of clinical translation, 124–125
 ethical concerns, 122–123
 for preclinical studies, 123–124
 for use of embryonic cells, 122
Ethics team, 94–95
Exosomes, 47–48
Experimental animal study, 130–131
 followup, 131
 induction of animal model, 130–131
Extracellular matrix (ECM), 10, 112

F

Facilities, GMP, 89–90

G

Gelatin coating, 144
Gene manipulation, 114–116, 122–123
Genetic modification of stem cells, 77–78
Glycoprotein 130 (gp130), 18
GMP. *See* Good manufacturing practice (GMP)
Good manufacturing practice (GMP), 87*t*, 88–94
 basic equipment for stem cell laboratory, 90–94
 equipment, 90
 facilities, 89–90
 processing and process control, 94
 supplies and reagents, 94
Good tissue practice (GTP), 87*t*, 88
 environmental control, 88
Growth factor, 18
GTP. *See* Good tissue practice (GTP)

H

Hair follicle (HF), 16
Hematopoiesis, 10
Hematopoietic stroma, 10
Hemopoietic stem cells (HSCs), 4, 7, 40–41, 57, 73–74, 128
 bone marrow, 63
 peripheral blood, 63
 sources, 74*t*
HF. *See* Hair follicle (HF)
Histone methylation, 42
Histone modification, 42
Homing, 29–32
 enhanced and directed, 30–31
 factors affecting, 77–79
 mechanism of, 30
 stem cell trafficking, 31
  mechanism of, 31–32
  in regeneration, 32

HSCs. *See* Hemopoietic stem cells (HSCs)
Hypoxia, 78

I

IGF-1. *See* Insulin-like growth factor I (IGF-1)
Immunocytochemistry, 129
Immunomodulatory functions, 51–52
Injection
  cell function after, 112
  frequency of, 101–103
  optimum number of cells for, 79–80
  routes of, 76–77, 77*t*, 99–101
Insulin-like growth factor I (IGF-1), 45
Intestinal epithelial niche, 16–17
Intestinal stem cell (ISC), 16
ISC. *See* Intestinal stem cell (ISC)
In vitro experiments, 128–130, 136
In vivo cell tracking, 109–110
  experimental, 109
  in humans, 110
Iron oxide nanoparticle labelling, 145
Iron oxide tagging, 103

L

Laboratory guidelines, 87–95
  personnel qualifications and job description, 94–95
  quality control team, 96
  structural facilities, 88–94
Laminin coating, 144–145
Lectin-positive progenitors, 63–64
Leukapheresis, 139–140
Lipoaspiration, 140
Local administration, 76
Long-term HSCs (LT-HSCs), 12

M

Macrophages, 52
Medical innovation, 125
Medical monitoring and adverse event reporting, 125
Mesenchymal stem cells (MSCs), 16, 40, 57–62, 73–74, 75*t*, 86, 101, 128
  adipose tissue, 61
  angiogenic potential of, 46
  bone marrow, 58
  cord tissue, 58–61
  culture, 142–143
  culture conditions, 46
  identification of, 143
  immunomodulatory function of, 51–52
  isolation from Wharton jelly, 141–142
  neural differentiation, 129*f*
  niche, 14–15
    in adipose tissue, 15–17
    bone marrow, 14–15
  peripheral blood, 58
  secretome, 49*t*
Messenger RNA (mRNA), 47
Microenvironment, 10
Microenvironment modulation, 51–52
  antiinflammatory effects, 51
  antimicrobial functions, 51
  immunomodulatory functions, 51–52
MicroRNAs (miRNAs), 42
Microvesicles, 47–48
Mitochondrial transfer, 49–51
  effects, 50
  inducers of transfer, 50
  mechanisms, 50
  variability in, 50–51
Mode of regenerative function, 112–113
Mogrify algorithm, 43
Molecular mechanisms, 41–42
Mononuclear cell separation, 140–141
MSCs. *See* Mesenchymal stem cells (MSCs)
Multifactorial (vitiligo), 127
  in vitro experiments, 128–130
  defects, 135–137
    clinical results, 137
    in vitro experiments, 136
Multipotent stem cells, 24

N

Neuroregeneration, 128–130
Niche, 7, 10–11
  in adipose tissue, 15–17
  bone marrow, 14–15
  common features of, 10–11
  epidermal stem cell, 16
  functions, 17
  intestinal epithelial, 16–17
  regulation of niche function, 17–19

NK cells, 52
Noncoding RNA regulation, 42
Nonembryonic adult pluripotent stem cells, 23
Notch signaling, 18

O

Oligopotent stem cells, 24
Optimal timing, 79
Optimum laboratory protocol, 74–76
  differentiation, 75
  expansion, 75
  purification, 74–75
Osteoblasts, 13
Osteons, 13
Oxazolone, 114

P

Paracrine action, 48–49
Parkinson's disease, 70, 72*t*, 114*t*, 128
Pericytes, 13, 15, 44
Periodontal ligament stem cells, 62–63
Peripheral blood hemopoietic cells (PB-HSCs), 62, 139–140
Plasticity of stem cells, 25–27
Plating cells for differentiation, 143–144
Pluripotent stem cells, 23
  artificially derived pluripotent stem cells, 23
  embryonic stem cells, 23
  nonembryonic adult pluripotent stem cells, 23
Potency, 22–24
  bipotent stem cells, 24
  cell cycle profile of pluripotency, 24
  molecular determinants of pluripotency, 23–24
  multipotent stem cells, 24
  oligopotent stem cells, 24
  pluripotent stem cells, 23
  totipotent stem cells, 22–23
  unipotent stem cells, 24
Preclinical, experimental, 112–117
  challenges in animal experiments, 117
  ethical considerations, 117
  small *versus* large animals, 116
  value of animal experiments, 112–116
Preconditioning of stem cells, 78
Proof-of-concept, 123–124
Purification, 74–75

Q

Quality control, 87*t*, 96

R

Reactive oxygen species (ROS), 50
Regeneration mechanism, 38–42
  contribution to tissue regeneration, 40
  transdifferentiation, 39–40
  variation in transdifferentiation potential, 40–41
Regulatory bodies, 86–87
Reservoir, 10–11
Routes of administration, 77*t*, 112

S

Secretome, 49*t*
Secretory function, 47–49
  exosomes, 47–48
  microvesicles, 47–48
  paracrine action, 48–49
Self-renewal, 25
  regulation of, 25
Signaling pathways, 42–43
Species selection, 113
Spinal cord injury, 72*t*, 80*t*
Somatic cell nuclear transfer (SCNT), 122–123
Standard operating procedures (SOPs), 85–86, 86*f*, 94–95
  coating of flasks, 144–145
  collection
    of bone marrow hemopoietic stem cells, 140
    of peripheral blood hemopoietic stem cells, 139–140
    of stem cells from adipose tissue, 140
    of stem cells from cord blood, 140
  differentiation
    evaluation of, 144
    plating cells for, 143–144
  isolation
    mesenchymal stem cell isolation, 141–142

mononuclear cell separation, 140–141
labeling, 145–146
mesenchymal stem cell culture, 142–143
identification of mesenchymal stem cells, 143
Stem cells, 3–4, 7
activation of endogenous repair system, 43–44
administrative team, 94
characteristics of ideal source of, 57
classification, 21–22
clinical evaluation team, 95
clinical followup team, 95
collection from adipose tissue, 140
collection team, 95
cord blood, 63–64, 140
definition, 21
differentiation, 29
engraftment, 29–32
ethics team, 94–95
handling team, 95
homing, 29–32
induction of angiogenesis, 44–46
mechanism of regeneration, 38–42
contribution to tissue regeneration, 40
transdifferentiation, 39–40
variation in transdifferentiation potential, 40–41
microenvironment modulation, 51–52
antiinflammatory effects, 51
antimicrobial functions, 51
immunomodulatory functions, 51–52
mitochondrial transfer, 49–51
effects, 50
inducers of transfer, 50
mechanisms, 50
variability in, 50–51
mobilization, 139
periodontal ligament stem cells, 62–63
properties, 22–29
evidence of stem cell transdifferentiation, 27
potency, 22–24
self-renewal, 25
stem cell plasticity, 25–27
theories of transdifferentiation, 27–29
regulation of tissue regeneration, 38
retention, 13
secretory function, 47–49
exosomes, 47–48
microvesicles, 47–48
paracrine action, 48–49
signaling pathways, 42–43
skin mesenchymal stem cells, 62
tissue-specific stem cells, 64–66
trafficking, 14, 31
mechanism, 31–32
in regeneration, 32
Stem cell therapy, 85, 105, 106*f*, 107, 121, 131
cell dose, 108–109
cell lineage, 71–73
cell purification level, 106
cell type, 74
hemopoietic stem cells, 74, 74*t*
mesenchymal stem cells, 74, 75*t*
cellular defects, 70–71
cellular loss, 128–130
choice of purification strategy, 106–107
clinical translational study, 132–134
combined defect, 127
experimental animal study, 130–131
followup, 131
induction of animal model, 130–131
factors affecting homing of stem cells, 77–79
mesenchymal stem cell homing, 78–79
methods to improve stem cell homing, 78
frequency of injection, 101–103
cells, 103
patient, 101–103
laboratory guidelines, 87–95
personnel qualifications and job description, 94–95
quality control team, 96
structural facilities, 88–94
method of evaluation, 80
multifactorial, 127
in vitro experiments, 128–130
multifactorial defects, 135–137

Stem cell therapy (*Continued*)
clinical results, 137
in vitro experiments, 136
optimal timing, 79
optimum laboratory protocol, 74–76
differentiation, 75
expansion, 75
purification, 74–75
optimum number of cells for injection, 79–80
optimum route of injection, 76–77, 77*t*
regulatory bodies, 86–87
replacement of defective protein, 134–135
replacement of deficient protein, 127
results, 135
standard operating procedures for
coating of flasks, 144–145
collection, 139–140
differentiation, 143–144
isolation, 140–142
labeling, 145–146
stem cell differentiation status, 107–108
stem cell injection, 99–101
route of injection, 99–101
structural defects, 71
timing of intervention, 99
tracking method, 109–110
in vivo cell tracking, 109–110
Stemness, 25
Sterile supplies, 94
Strain selection, 113
Stromal vascular fraction (SVF), 61
Structural defects, 71
Supplies and reagents, 94, 146
Sympathetic nervous system, 18
Syncytium, 48–49
Systemic injection, 76–77

T
T cells, 24, 52
Therapeutic efficacy, 112
Tissue regeneration, regulation of, 38
Tissue-specific stem cells, 64–66
Totipotent stem cells, 22–23
Tracking method, 109–110
in vivo cell tracking, 109–110
Transdifferentiation, 39–40
evaluation of, 29
evidence of stem cell, 27
factors affecting frequency of, 28
induction of, 28–29
molecular mechanisms of, 41–42
requirements for proof of, 28
theories of, 27–29
Transforming growth factor beta (TGFB) pathway, 42–43
Translational stem cell research, phases of
clinical trials, 117–118
laboratory experiments, 111–112
preclinical, experimental, 112–117
challenges in animal experiments, 117
ethical considerations, 117
small *versus* large animals, 116
value of animal experiments, 112–116
Trinitrobenzenesulfonic acid, 114

U
Unipotent stem cells, 24

V
Vascular endothelial growth factor (VEGF), 15, 45
Vitiligo, 127, 135–137
Voluntary informed consent, 124–125

W
Wharton jelly, 58–62, 141–142

FSC
www.fsc.org
MIX
Papier | Fördert gute Waldnutzung
FSC® C083411

Kolibri 360 GmbH
Ferdinand-Jühlke-Straße 7
99095 Erfurt, Deutschland
produktsicherheit@kolibri360.de